The Melencolia Manifesto

The Melencolia Manifesto

David Ritz Finkelstein
*School of Physics, Georgia Institute of Technology**

* For further communication contact Shlomit Ritz Finkelstein <finkelstein@emory.edu> or Aria Ritz Finkelstein <ariaritz@mit.edu>.

Morgan & Claypool Publishers

ISBN 978-1-6817-4090-4 (ebook)
ISBN 978-1-6817-4026-3 (print)
ISBN 978-1-6817-4218-2 (mobi)

DOI 10.1088/978-1-6817-4090-4

Version: 20161201

IOP Concise Physics
ISSN 2053-2571 (online)
ISSN 2054-7307 (print)

A Morgan & Claypool publication as part of IOP Concise Physics
Published by Morgan & Claypool Publishers, 40 Oak Drive, San Rafael, CA, 94903 USA

IOP Publishing, Temple Circus, Temple Way, Bristol BS1 6HG, UK

For five years or so David had a mistress, Melencolia. After dinner, he would withdraw to our library where we had copies of two versions of the engraving at two different magnifications. He would reflect in front of one of them or switch between them for some time, or for the entire rest of the evening. David shared his affair with me, and Melencolia, Dürer's unique contribution to Renaissance art and science, has become a member of our family.

This manuscript is David's, neither mine nor his children's. But we all dedicate it to his memory.

Shlomit Ritz Finkelstein

When my brother Daniel, my sister Eve, and I were growing up, our father's interest in creation extended to household objects around us. While seeking a unified field theory, he might pick up a breakfast spoon, turn it in the air, and reflect on what its design revealed about its manufacture. He read anything he got his hands on, from the Oxford English Dictionary to lists of ingredients on cereal boxes. That undying (though we lost him, it remains) curiosity encounters its natural partner in Dürer's engraving. As an adult, I had the honor to work with my physicist father just once professionally, that is, in editing his early essay on MELENCOLIA§I for publication in The Saint Ann's Review. With this new publication of his more extensive study of the subject, that honor seems all the more acute.

Beth Bosworth

Contents

Author biography

David Ritz Finkelstein (July 19, 1929–January 24, 2016)

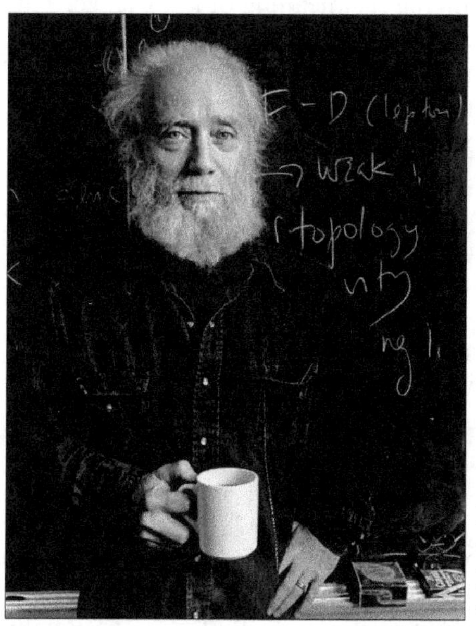

The last and longest phase in Finkelstein's scientific life is best described by him as spent working on 'reconciling the fundamental concepts and principles of quantum theory and space-time theory, Einstein and Heisenberg.' He used quantum set theory, his extension of classical set theory, to do so and presented the then state of his work in his 1996 book *Quantum Relativity: A Synthesis of the Ideas of Einstein and Heisenberg*. He continued to work on the fundamental question of the structure of space-time for the rest of his life.

Throughout his career Finkelstein published over 100 scientific articles, some of them seminal. In 1958, as a young man, he published his paper about the 'unidirectional membrane,' which would be renamed five years later the 'black hole.' He was the first to discover kinks, topological charges and topological spin-statistics theorems, with Misner (1959) and Rubenstein (1962). The 1962–63 papers with Jauch, Schiminovich and Speiser were the first to introduce the Higgs mechanism before Higgs. Inspired by John von Neumann, Finkelstein was a pioneer in considering logics as physical. He understood the peculiarities of quantum physics as demonstrating quantum logics, which unlike classical Boolean logics is non-commutative. Later in his life he also explored the implications of non-associativity. Between 1969 and 1972 Finkelstein published his series 'Space-time code' in which he considered the universe as a quantum computer and studied the fundamental unit of time, which he named the 'chronon.' He also was a pioneering voice when he argued the feasibility of quantum computing in 1969.

He earned his PhD in theoretical physics from the Massachusetts Institute of Technology in 1953. He taught at Stevens Institute of Technology, was a professor of physics and department chair at Yeshiva University, and after more than twenty years of service retired as an emeritus professor of physics at the Georgia Institute of Technology. For over two decades, Finkelstein edited the *International Journal of Theoretical Physics*.

He is survived by his first wife Helene Cooper and their three children Dan Finkelstein, Beth Bosworth, and Eve Finkelstein, and by his second wife Shlomit Ritz Finkelstein and their daughter Aria Ritz Finkelstein.

Foreword

My father was a quantum physicist, not an art historian. What in Dürer's engraving so captivated him? He contextualized his own work within a long history of scientific thinking, within a long legacy of thinkers changing the meaning of knowledge. This historical view meant that, where Newtonian physics was the way to understand God's rules for the world, modern quantum physics challenged the very existence of such rules and therefore the very nature of knowledge. Following Francis Bacon's understanding that scientific thought often reveres false 'idols,' my father worked to identify and then smash the idols of physics. In an unfinished manuscript, he explains quantum theory as 'the theory of what it means to know.' He writes,

> In quantum physics we no longer imagine that it is possible to describe any physical process completely and precisely, as (say) Descartes imagined. Instead we describe the most detailed possible input and output actions of experimenters and give the probability that a given input will be followed by a given output. This input-output philosophy makes quantum theory more like a cook-book or a computer-programming manual than a traditional book on celestial mechanics.

Even before the hidden faces in Melencolia§I began to emerge, my father was pulled into the engraving as an exploration of what it means to know. One never sees the back of the Dürer solid. 'The Octahedron is a puzzle that is intentionally unsolvable,' he writes. '[Dürer's octahedron] declares that the Intellectual World has a mathematical design, but that this design is inaccessible to us.' In his youth my father would have agreed, that it is not accessible to us *yet*. But what he learned from quantum physics was that we don't look through the language to find its meaning, through the picture plane to find the solid, or through numbers to find the physical truth.

For a time, whenever I saw my father he would announce a new face in *Melencolia§I* and ask me if I could see it or not. Was it really there? His rubric was that if he couldn't find three justifications or explanations for a potential face to exist, it didn't. Without such a rule, once you start hunting for hidden faces, you start finding them all over the place. Dürer's engravings in particular: there is something about how they are drawn, maybe an excess of detail or texture, that makes you feel certain that if you look at them correctly the face will emerge. Occasionally it does, but often it doesn't. It's just that there's a face-ness to Dürer's naturalistic style. My father and I wondered about this quality, and it occurs to me now that it is in the spirit of the Dürer solid. A hidden face is a ripple on the surface of the picture plane. One cannot see both the Dürer solid and Dürer's mother at the same time. Likewise, the emerging face disrupts one's visual world and replaces it with another, drawing attention to the engraving as an object.

While Dürer found melancholy in his solid, this 'subtler, more self-conscious epistemology' of quantum theory was my father's joy. The quantum epistemology, in his own words:

> This is the kind of section I would skip when I was starting out in physics. I wanted hard mathematical facts, not philosophical chit-chat. 'Number is the language of science,' Pythagoras tells us across the centuries. Number, not words. Is that not a thrilling idea? Mathematics, an almost secret language that Man has created in just the last few thousand years, completely captures all of nature! Are we not like gods?
>
> Pay attention to that feeling. It is hubris, the arrogance that precedes disaster. For some it blocked the understanding of relativity, and now it blocks the understanding of quantum theory for many.
>
> Mathematics is not the language of science.
>
> Eavesdrop in a science laboratory. Are they speaking in mathematics? Are there mathematical terms for funding salaries and equipment, blowing glass, purifying silicon, swinging pendulums, publishing? Physics is a social science, as [Victor] Weisskopf used to say. Mathematics refers only to mathematical experience, a small part of physical experience, though an indispensable part.
>
> Quanta are so different from symbols, including mathematical ones, that no exact map of one into the other is possible. The key difference is that our observations of mathematical objects are usually complete while our observations of physical ones are never. Predicates about a mathematical system are assumed to commute, while physical observations are observed not to commute. Non-commutativity describes how observation changes the observed system, and this kind of change is ignored in the works of fiction that make up mathematics: 'In a far-away land lives an intelligence that can add 1 to 0 any number of times without end.' The quantum non-commutativity is experimental; the Boolean commutativity is axiomatic.
>
> That quanta have no exact descriptions does not mean that quanta do not exist. We live our daily lives with no exact descriptions of anything.
>
> Theoretical physicists left normal thought behind in the heyday of classical mechanics. Some natural philosophers thought that they made complete mathematical models of nature, or at least could do so in principle. In quantum theory, physicists return to the human predicament.
>
> Another example of such a radical change in physical thought is the special theory of relativity. This relativized our concept of 'the present.' Quantum theory does much the same for our concept of 'the state of the system.' Quantum theory is the most successful physical theory in history. Nevertheless, much confusion envelops it. Many who practice it most productively claim that no one understands it, certainly not themselves.
>
> Much of this confusion arises from deep, ancient, and usually tacit magical beliefs about how one knows anything about natural objects, ideas that conflict with how we actually learn about Nature.

The process by which we see the world around us is invisible to our ordinary senses, as though it were magical. One old idea is that we just know, as by direct contact between knowing spirit and known object, or as one knows the Logos according to Parmenides. The great natural philosopher and mathematician Leibniz proposed that God preordained this puzzling harmony between the world and our mental images of the world.

Also old, however, is the alternative notion that we see an apple when small particles of light fly from it to our eyes. We did not see them in flight or perceive the intricate neural processes that they trigger when they reach our eyes. We experience only the results. Nevertheless, perception is not magic but a chain of subtle causes and effects, in which the apple, light, and our mental image of the apple are all equally physical.[1]

Aria Ritz Finkelstein

[1] Finkelstein D 2013 *The Idea of the Quantum* unpublished manuscript.

Editor's note

The following manuscript is one that David Ritz Finkelstein began working on in the early 2000's. It went through several drafts, an early one of which was published in 2004 as an article in *The Saint Ann's Review*. The spirit of the work and its conclusions have changed little, but its order and several of its specific insights have evolved. Some of the earlier iterations flow more smoothly, but this manuscript reflects Finkelstein's latest thinking. Accordingly, it displays some of the jagged seams of its construction, as befits an ongoing and thus imperfect conversation.

IOP Concise Physics

The Melencolia Manifesto

David Ritz Finkelstein

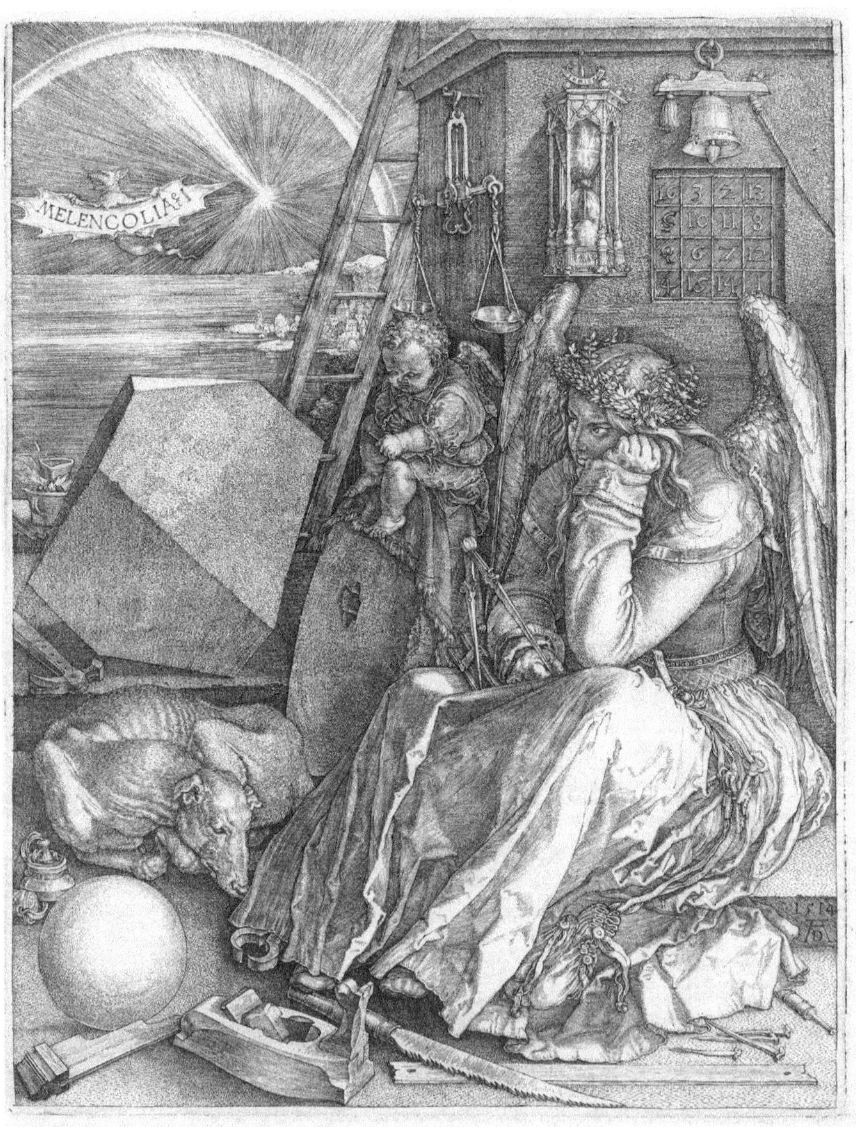

Figure 1.1. MELENCOLIA§Ia, Albrecht Dürer, 1514. https://commons.wikimedia.org/wiki/File:Albrecht_D%C3%BCrer_-_Melencolia_I_-_Google_Art_Project_(_AGDdr3EHmNGyA).jpg

An earlier version of this book was published in 2004 as a paper in *The Saint Ann's Review*.

doi:10.1088/978-1-6817-4090-4ch1

1. Origins

In 1455 Muhammad II, conqueror of Constantinople and founder of the Ottoman empire, invaded Hungary in response to Crusader invasions, and one Albrecht Dürer, 23, left the village of Ajtás (or Eytas) in Hungary and eventually resettled in Nuremberg. *Ajto* is Hungarian for door, *Thür* in the German of the time, so Albrecht of Ajtás became Albrecht Thürer; which can also mean 'Albert the door-maker'. He soon simplified the spelling to Dürer, a homonym in one dialect.

Albrecht married his master's daughter, Barbara Holper, and they had 18 children. Three lived, the brothers Endres (Andrew), Albrecht, and Ajtás (Hanns). Albrecht the younger became the most famous engraver of the North German Renaissance, although he remained a Hungarian national in Nuremberg all his life. One of his most famous engravings, MELENCOLIA§I, contains a secret manifesto (to coin an oxymoron) that has escaped full attention until now.

2. Background

Albrecht Dürer (1471–1528) hid some of his most deeply held beliefs in full view in the engraving usually known as MELENCOLIA I (1514) (figure 1.1). As a work of art it influenced European art for centuries [2], and as a puzzle it has challenged many riddlers before us. Details have been overlooked in recent centuries, however, and its meaning, once known but apparently repressed, is in the details. Among other things, this rich work is a carefully constructed puzzle whose pieces are to be discovered and decoded. I deal here with about thirty such clues encrypted in the fine structure of the engraving. While I am reasonably sure of the resulting exegesis as far as it goes, I am surer of the underlying images, so I list them below as a challenge for the next and better interpretation. I have found two pointed out in the literature: the first Ghost [24] and the Flood [16]. Any mistakes in my own interpretation should cast no doubt on these rediscoveries. The point is not what I see in the engraving but what Dürer expected his viewers to see. The philosophical views I read in the engraving and present below are not necessarily my own. They are selected from or extend those expressed in writing by Dürer and his acquaintances.

Any casual observer of MELENCOLIA I will note the impressive angel sitting in the right foreground and the small winged serpent flying in the left background, like virtue and evil personified. A winged boy, or putto, a familiar convention of Renaissance art, perches between them on an upturned millstone and intently inscribes a block. Paraphernalia of the arts and sciences clutter the scene in a significant pattern.

The art historian Erwin Panofsky [2] saw the engraving as intensely autobiographical and read its melancholy as Dürer's own frustration at the gap between artistic and divine creation. Frances Yates, historian of the Hermetic tradition, took its melancholia to be one that inspired creativity, not mere depression, and read the engraving as a declaration of the harmony between microcosm and macrocosm: as above, so below [3]. The art historian Patrick Doorly sees it as an illustration for Plato's Greater Hippias, a dialogue on beauty [4], with the angel melancholy because the artist cannot define absolute beauty. Such critics pass over more than a score of individual concealments and paradoxes in the engraving: a key anagram, rebuses, Cabalistic gematria, puns, a Star of David, over a dozen hidden faces, two Jews blowing the ram's horn, the entire figure of a kneeling warrior in Arab dress, and even the downfall of Babylon. There are references to *Genesis*, *Revelation*, Cabala, the Nicean Creed, and the Dürer family history. There are numerous demonstrations of the limits to human perception. These illuminate Dürer's views on art, science, philosophy, and religion, and support some of the interpretations of others that I have just cited, but not all. The interpretations I cite do not mention what may be the main point of the engraving.

That point concerns relativity in the broad sense, the study of how what we see depends on our viewpoint. Dürer uses perspective to construct several examples of relativity and ambiguity. It is not anachronistic to attribute this idea to Dürer, since he is one of the creators of descriptive geometry, which is precisely a theory of how the same object looks from different points of view, and is probably the first mathematical theory of relativity in the broad sense. He is significant in the history of the perspective technique, which the Romans and Greeks practiced and the Church suppressed as deceptive. Dürer carried 'the secret of perspective,' as he called it, from Italy to northern Europe, and some say that his book on the subject founded the mathematical theory of descriptive or projective geometry, though this developed over centuries with growing rigor. Perspective, special relativity, and general relativity all rest on an understanding of visual perception and light, and so there are family resemblances among them.

The apparent foreshortening of limbs in perspective and the apparent shortening of rulers and slowing of clocks in relative motion or in gravitational valleys are all effects of relativity in the broad sense. Where relativists today speak of light-cones, Leonardo and Dürer spoke of 'pyramids of vision,' which are light-cones with time omitted, being made of paths of light in space rather than trajectories in space-time. They held that in perception one such pyramid diverges from the point observed, like a future light cone, and another converges to the observing eye, like a past light cone. The perspective transformations of the graphic plane that we carry out when we shift our viewpoint are special cases of the transformations of space-time used later in modern relativity. Both are represented by matrices or tables of two rows and columns, with real or complex number entries respectively. Projective geometry is a playground for relativists and made Dürer an eminent mathematician of his time.

There are, to be sure, well known limits to Dürer's knowledge. Most famously, the astronomer Kepler pointed out that in Dürer's work on measurement [18] he drew the ellipse produced by cutting a cone with a plane as egg-shaped, with its small end near the tip of the cone. I found a striking error in optics in this engraving. Nevertheless, to catch his meanings we must not underestimate his power with the burin, nor his acute observations of nature. We should remember that Dürer was a humanist, knowledgeable in mathematics, poetry, and antiquity as well as art, and that he argued philosophy hotly and on equal terms with a famous scholar like

Pirckheimer and conversed with Erasmus. We should be aware of his impassioned criticism of corrupt ecclesiastical practices and his support of the Reformation.

Like Leonardo, Dürer began his training under a goldsmith, namely his father, a Hungarian immigrant in Nuremberg. By the year on the engraving, he had made two historic visits to Venice, where he studied perspective and human proportion and developed a passion for geometry. Eventually he extended Leonardo's work on beauty in a relativistic direction, using projective transformations to allow for different standards of beauty.

Dürer was the first to publish in German a mathematical proof or a book on pure mathematics. He produced the first printed star-chart and the first printed map of the world as a sphere viewed from outer space. He shaped the German scientific language much as Leonardo did the Italian: by describing the processes of art and nature in ordinary language instead of flowery Latin, painstakingly picking out the prose ornaments and classical references that infested scholarly writing in his time.

He was a brilliant and recognized originator in art too. He did the first known self-portrait and the first known specific landscape. He invented etching and some perspective drawing machines. He seems to have invented the modern artist, being the first to charge for his special talent instead of just billing for time and materials like Raphael or any house painter of the time. His life mission seems to have been to sanctify art above artisanry and even above philosophy; to elevate graphic art to the status of music. Number lifted the medieval quadrivium of arithmetic, geometry, astronomy, and music out of the mundane and into the divine. So Leonardo, Dürer, and others sought to found art and beauty too on number. The concept of the Golden Rectangle, Leonardo's famous drawing of a man inscribed in a circle, and Dürer's books on descriptive geometry and body proportions are likewise mementos of that search. So is much of MELENCOLIA§I.

3. Gateways

A demon carries a banderole that bears the inscription 'MELENCOLIA§I' (see figure 1.1). The symbol §, a stylized S, is usually taken to be a meaningless ornamental flourish and omitted from the motto and the name of the engraving. Some also correct the spelling of 'Melencolia'. These corrections erase the main clue to the meaning of the engraving. My decryption is subject to a warning that Dürer encrypted into the engraving, exemplified in the engraving itself, and left in writing as well:

> Our senses deceive us, we do not see things as they are, we make uncertain reconstructions and do not even realize that we do so.

To minimize my projection of modern views onto Dürer's, I base this decryption on works that he knew well:

- His own works, writings, and coat-of-arms.
- The works of Leonardo da Vinci, who worked on the *Mona Lisa* in Florence when Dürer studied 'the secret of perspective' in Venice.
- The Jewish and Christian Bibles.

- The *De Occulta Philosophia* of Heinrich Cornelius Agrippa von Nettesheim (1486–1535), who carried it to Nuremberg shortly before this engraving [5].
- The *Hieroglyphica* of Horapollo [17], which Dürer illustrated and then used as key to a major allegorical work of the preceding year.

3.1 Ghosts

I admired MELENCOLIA§I for many years but I was first drawn into decrypting it when it flashed across my computer screen with an outline of a subliminal face that I had missed for decades. Step back several paces from a good print and look at the shading of the front face of the Dürer octahedron for a minute or so, and you too will probably see a hint of a woman's face in profile. I call her First Ghost, because, as we will see, she inhabits Sphere III of the angels. This is one of many subliminal images in this engraving. Some are important for the interpretation that I suggest.

One calls such images subliminal because they wait just below the *limen* or threshold of casual perception. 'Limen' is itself a subliminal reference to Dürer, which can mean 'door-maker'. Dürer put a gateway (limen) into his coat-of-arms. Gates and doors and limens dominate MELENCOLIA§I, such as the limen in the sky formed by the Moonbow.

For some this subliminal face is a frightening Death's head [24]. Dürer's last portrait of his mother was done in the same year, the year of her death. Some see it as malicious, but it can also be seen as a lovingly truthful depiction of the physical consequences of bearing 18 children, and of the familial amblyopia, known as the *Dürerblick*. Perhaps First Ghost is his mother in Heaven, for the octahedron will turn out to be the world of angels.

3.2 Anagram

Alerted by the subliminal First Ghost, I turned my attention to the most conspicuous encoding in MELENCOLIA§I, its motto.

Melancholia is transliterated Greek for melancholy, which originally meant black bile. One Latin form is *melancolicus*, the German is *Melancholie*. The Italian *malinconia*. Dürer himself uses the etymologically reasonable form *Melancolia* in a title-page woodcut of 1503. In no language do I find the word *melencolia*; the spelling is odd—even for Dürer. The spelling of the time was variable, but this spelling and the appended 'I' suggest a message that Dürer wished both to record and conceal.

Agrippa was a prolific and controversial figure of the Florentine Neoplatonist school [5]. Anagrams and gematria are found in his *Occult Philosophy*. They are also among the techniques used to squeeze new meanings out of old texts in the *Baraita of Thirty-Two Rules* of Rabbi Eliezer ben Jose the Galilean, who flourished in the second half of the 2nd century CE and was cited importantly in later years. Such lore became accessible to Florentine Neoplatonists, including Agrippa, an active Hebraist, when some of the Jews expelled from Spain in 1492 settled in Florence.

Whatever the source, by the 16th century letter-permutation was a standard way to protect intellectual property. The Royal Society and the system of scientific archives were still over a century in the future. Writers of Dürer's time who solved an important problem could not guard their intellectual property by publishing or patenting it. Some did so by scrambling its letters and publishing them with the problem. Earlier Roger Bacon had scrambled the formula for gunpowder in this way in order to prevent the proliferation of this terrible weapon. Later Galileo would scramble his discoveries of the phases of Venus and the rings of Saturn. Mathematicians scrambled their theorems. The resulting meaningless jumble of letters declared on its face that it was a cipher. Sometimes the hidden message was misread by a false rearrangement.

In the next degree of concealment, the scrambled letters themselves spell out a cover message, the anagram. This can hide not only the content of the message but even its existence. MELENCOLIA§I has done this well. Only the cover message has been read in recent centuries as far as I know. I unscrambled the anagram as soon as I began to interpret the engraving, as follows.

Since Panofsky declared the engraving autobiographical, I tentatively supposed that the motto referred to Dürer himself. To see how he might describe himself graphically I went to his coat-of-arms (1490, 1523), figure 1.2.

His famous monogram and the year are top center. Below them come clichés to be found in any dictionary of heraldry [25]: a blackamoor for heroic action by an ancestor in the Crusades; eagle wings for fame and glory; a closed helmet in full profile for an esquire, the lowest estate. A shield below them bears his familial coat-of-arms:

- The coat-of-arms shows a *gateway with open doors*. This is an ideograph for both 'Dürer', which means both 'man of the door', and 'man of Ajtás', the town in Hungary from which the family came to Nuremberg.

This sent me looking for other ideographs in the coat-of-arms.

- The gateway stands on a cloud, more prominent in the 1490 version than the 1523.

I read the gateway as follows. Coats-of-arms usually display the pride of the family, some great personal accomplishment. To immortalize his talent for engraving, Dürer could simply have drawn a burin on the shield, or the word 'burin' itself. The Latin for burin is *caelum*, which is also the ordinary word for 'the sky', 'Heaven', and 'the heavens'. It has the same root as our 'celestial'. *Caelo* means both 'to Heaven' and 'I engrave'. The root seems to mean swelling or hollow; both the sky and a burin are hollowed out.

- The Dürer coat-of-arms is both an ideogram for 'gateway to Heaven' and a rebus for 'the burin (is) the gateway'.

'Gateway to Heaven' was already a well-known metaphor for the Roman Church itself. A person of the time might take the Dürer coat-of-arms not as a sacrilegious display of ego but as a pious reaffirmation that we are made in the image of God. But it can also serve as a banner of humanism: art too is a gateway to Heaven.

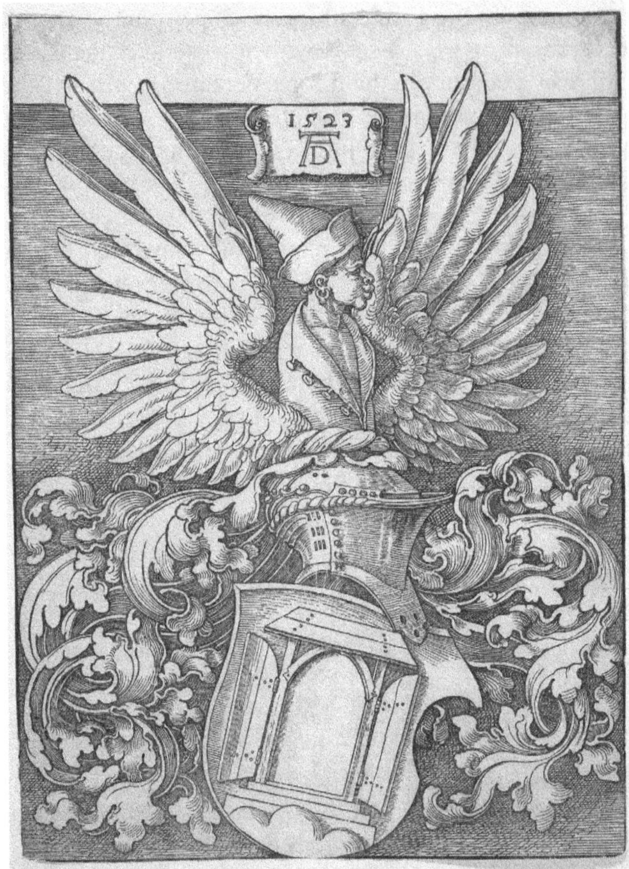

Figure 1.2. Dürer's coat of arms, 1523.

Gates of Heaven open again and again in MELENCOLIA§I itself. The Dürer monogram AD, exhibited at the top center of the 1523 coat-of-arms and slightly above the bottom right corner of MELENCOLIA§I, is one of them. It seems at first to be a pun on *Anno Domini.* Now one can see that it is also an ideogram. The flat-topped A is a gateway as well as a letter. The legs of the A are the uprights of the gateway. The lintel sits on them and connects them. A stiffener just beneath the lintel is the crossing of the A, rather high for an A but proper for a gateway. The D between the uprights of the A echoes the doors between the uprights of the gateway in the Dürer coat-of-arms.

In Dürer's coat-of-arms he symbolized his heavenly art by the *caelo* rebus. Once I had found this pun I checked to see whether CAELO fits into MELENCOLIA. It does.

The leftover letters then spell out LIMEN, commonly meaning gateway, a self-reference to 'Dürer', and a good description of the gateway in the Dürer coat-of-arms. Limen also means gate, doorway, threshold, lintel, walls, house, home, boundary path, and limit, depending on context.

There remains the '§I' of the banderole. Some suggest that 'I' refers to one third of the medieval tripartite universe, and expect that MELENCOLIA II and III were engraved and lost, or planned and never executed. MELENCOLIA§I is indeed one of three but they all survived famously well; together they are called the Master Engravings of Dürer, dated within a year of one another. The other two are *The Knight, Death, and the Devil* and *St Jerome in His Study*. MELENCOLIA§I, *The Knight*, and *Jerome* exemplify Faculties I, II, and III of the Neoplatonic triad of the natural, political, and theological philosophies. Agrippa's Book III, the Hebrew Bible, lies open before Jerome as his Book I, Nature, lies open before the boy. Likely the knight as a statesman represents Faculty II. We see why one engraving bore its own caption and not the other two: Dürer proselytizes for Gateway I, not as one of three paintings but as one of three gateways. And this supports the 'I' interpretation of Boorsch and Orenstein [6].

- The full motto 'MELENCOLIA§I' is an anagram of 'LIMEN CAELO §I', meaning both 'the holy first gateway to Heaven' and 'the burin (is) holy Gateway I', if Dürer makes the same pun on *caelo* in the motto that he does in his coat-of-arms.

Another usage of 'I' would also have been natural for a mathematician like Dürer. Even in 1514, an '*x primus*' in a mathematical writing referred back to an earlier *x* and did not imply a later *x secundus* or *x tertius*. This is true for Dürer's own mathematics too. Then the motto would mean, however, that Dürer was superseding the old gateway to Heaven with a new one, thus claiming innovation. This was a capital crime and so would explain the secrecy; but the accusation does not hold water. According to Dürer's writings, he did not consider himself an innovator but a renovator, stripping away the innovations made by the Church; a basic humanist position.

Panofsky was right, and we have another signature. This gave me some confidence that I was conversing with Dürer, not just with myself. The unscrambled motto combines the family name-rebus, Dürer's art, his life goal, and his coat-of-arms itself. It supports the interpretation that Dürer intended the clouds in the coat-of-arms as a symbol of Heaven. The hidden phrase also applies to the dim archway in the heavens that frames the motto, and to other elements that we take up later.

We need not postulate lost engravings of the three worlds to explain the motto. MELENCOLIA§I itself shows all three Neoplatonic faculties, worlds, and gateways to Heaven. The boy holding the burin, presumably Dürer himself, is Agrippa's imaginative Faculty I choosing Agrippa's Gateway I, the natural philosophy that evolved into physics. For the artist-scientist Dürer, natural philosophy meant making precise and beautiful records of Nature and knowing and obeying the laws of perspective and structure. The modern chasm between art and science did not exist for him. Gateway II is the way of mathematical philosophy and the Prophet, who, by elimination, must somehow be represented by the dog.

A dog? Can this be right? I will come back to this later.

Gateway III, the way of theological philosophy, is the contemplative way of the angel.[1]

3.3 Subliminal faces

The subliminal faces in the upper pentagonal face of the octahedron I call the ghosts. Which we see depends on how we turn our heads relative to the print, for our acute innate ability to find faces is sharply tuned to upright ones.

- First Ghost, well known, is the woman in profile with her head erect, already mentioned.
- Second Ghost is a younger bearded man seen in full profile looking to our left with his head upright, to the left of First Ghost.

Other subliminal faces lurk in this engraving. We have to scan for these subliminal faces, also called secondary images, with some persistence before they emerge. They are subtle, ambiguous, and strain our perception. Do we see them or create them? Most of the left edge of the face of First Ghost has almost no physical correlate in the engraving. Dürer compels us to see it, when we look for it, by supplying chin and forehead lines and eye and nostril shadows, but we ourselves create the rest of the face. These subliminal faces are ambiguous, lie at the edge of perception, and are eminently deniable. They are joint productions of the hand of the artist and the mind of the viewer, demonstrating Dürer's mastery of both media. When we see one of the faces in this quaternity we do not see the other three, mainly because they share some elements. For example, Second Ghost and First Ghost share an eye-pit.

We are remarkably apt at imagining faces where none are intended, as in the Moon and in maps of the ocean floor. Subliminal faces are common in Dürer's work, but be aware that scanning, digitization, or lithography creates accidental ones and distorts or destroys those already there. In an earlier version of this book, I once claimed that in Dürer's drawing *Orfe der erst Pusaner* there was a woman's face in the treetop, based on a lithograph. In better prints with no perceptible grain this lady vanishes.

So I have dutifully tried to entertain the possibility that the subliminal faces in MELENCOLIA§I are imaginary, but it seems that Dürer made this difficult by injecting so many. We see him rehearsing subliminal faces in his 1493 drawing *Sechs Kissen* or *Six Cushions*, where each cushion has at least one grotesque face.

[1] In a previous draft (2006), this three-world cosmology structured Finkelstein's interpretation of the engraving. He organized the essay's sections according to the three world structure, from third to first. Under 'Third World' he included sections titled Four Ghosts, Octahedron, Magic Squares, Another 'Melencolia I,' Ladder, House, Door, Angel, Fools, Purse and Keys, Compasses, and Four Prophets; under 'The Second World' he included sections titled Millstone, Dog, Flood, and Destroyers; and, under 'The First World' he included sections titled Limen caelo, Boy, Four Friends, Globe, Moonbow, Light, Hexagram, Bellringer, Melancholy, Skull, Unitarianism, Flourish, The serpents, and Summation. Here, Finkelstein will return to his discussion of the three worlds in section 3.6, The Demon.

Dürer's five subliminal quadruples go beyond the familiar kind of relativity, the change in visual appearance under rotation that turns ↑ into ↓. Each face appears and disappears as we merely rotate it about the line of sight; to see one at all we have to be in the right reference frame. Dürer hides some of them as paper makers hide watermarks. Our perception adjusts itself to the largest range of brightness that is present. We filter out distracting low-contrast background in brightness within a high-contrast image.

Viewers of Dürer's time and place may well have seen these faces easily and quickly, alerted by the many subliminal faces in earlier works of Dürer, and interpreted them at once. We mention some precedents to provide practice in face detection.

The most spectacular demonstration of Dürer's wizardry with illusion is his engraving variously called *Der Spaziergang,* or *The Lady and the Gentleman,* or *Young Couple Threatened by Death.* A brilliant subliminal image of a male face forms out of and looms over the young couple, on a larger scale than the overt engraving.

Their heads form his two eyes, her arm his mouth. The gentleman's plumed hat becomes a flaring eyebrow of the subliminal face. In this picture Dürer does not use the watermark technique but a kind of bricolage. Several overt forms have to be broken into pieces by our perception and reassembled to make the covert one. Also:

- His watercolor *View of Arco* has a cliff that scowls famously.
- His drawing of the same Arco hides three such angry faces in the landscape.

There are many such faces hiding in MELENCOLIA§I. This is a curiously domestic angel, dressed as a housewife, complete with keys and purse hanging from her belt. The keys and purse of the angel are the only elements in the engraving whose meaning has been given us by Dürer. They stand for power and wealth. But while his Madonna and his Venus wear their purses and keys at the waist, the angel drags her purse on the ground. Likely this means that theology is soiled by fiscal power, for Dürer was a passionate critic of clerical corruption. Only a fool would worship her, the artist seems to say.

And there the fool is, a crude low-browed subliminal caricature lurking near her feet in the hem of her gown, next to Dürer's monogram. Call him First Fool.

- We must turn the engraving about 60° counterclockwise to see First Fool best and then he looks to our right. His nose forms part of the bottom edge of the angel's robe. He has a mustache and curly hair.

A powerful winged housewife with a fool in her hem also resides in a woodcut attributed to Dürer in chapter 13, 'On Amours' (Von Buolschaft) in the *Ship of Fools* of Brant [7]. There the winged housewife is Venus, the goddess of the fools of love, the most common variety of fool according to Brant. Venus's eagle wings signify fame. We know she is a housewife because Brant calls her sarcastically 'Frau Venus'. The caricature in her hem is crude, but since it recurs it is probably part of Dürer's graphic language. Judging by her purse, keys, and what seems to be a spindle dangling from her left knee, the angel too is a housewife. Dürer seems to say that the theological philosophy also attracts fools.

Are the quaternities of the worships[2] and the ghosts echoed here? Let us look for more fools.

- It is easy to see the larger face of Second Fool looking at us and slightly to our left. Second Fool shares curled hair with First Fool.
- Third Fool is looking upward and to our left and has a body, a complete arm with a mitten-like hand at his side, a leg, and the purse of the angel.
- Fourth Fool stands upright to the left of the others, wearing a robe from head to ground, looking almost right at us.
- If we cock our heads about 30° to our left, we can see or imagine an erect face peering from behind the boy, in the part of the blouse billowing out between two straps. I call him First Friend. His head is bounded on our right by the curve of the boy's gown between the straps and is looking downward towards the viewer's left. The boy's gown, where it passes over the millstone, forms two near right-angles that look like a small book, and First Friend seems to have a hand on it. That and the robe trailing behind him make First Friend look clerical or at least scholarly.

First Friend too is part of a possible quaternity in MELENCOLIA§Ia that Dürer substantially reworks in MELENCOLIA§Ib.[3]

- To see the profile of Second Friend we should stand the engraving on its left edge. The fringe of the boy's garment then becomes the profile of Second Friend.
- Third Friend is the boy's left upper arm.
- In MELENCOLIA§Ia Fourth Friend is the boy's right lower leg.
- In MELENCOLIA§Ib Fourth Friend is a smaller better-formed head set into the boy's right knee, touching the boy's block.

In *The Knight, Death, and the Devil,* Dürer portrays the knight as spectacularly oblivious to Death and the Devil, though in other such encounters in the work of Dürer, the knight engages evil in mortal combat. The knight is generally said to represent the ideal Christian soldier described by Erasmus (c. 1466–1536), so virtuous that he was blind to the evil about him. Erasmus was a revered acquaintance of Dürer.

On the other hand, Dürer wrote a fervent plea to Erasmus, perhaps unsent, to join with him and Luther in battle against the 'Cave of Hell' and die a martyr's death. Erasmus remained within the Catholic Church and worked for reform instead

[2] Editor's note: Finkelstein worked to elucidate a quaternity of the four worships, represented by a set of subliminal figures in the angel's skirt. The analyses of these figures take slightly different forms in different drafts of this manuscript. Generally, First Worship is represented by Moses, a stereotypically Jewish man with a shofar. Second Worship is Jesus. Third Worship is Muhammad, represented by the Arab. The Fourth Worship, Finkelstein suggests (2006), might be a monotheistic prophetess. The addition of this group of subliminal faces gives us a quadruple of quaternities (four ghosts, four fools, four friends, and four worships or prophets), like the 4 × 4 table of the magic square, possibly derived from Agrippa's writing on the cosmical significance of the number 4.

[3] Editor note: In the section 'The angel' below, Finkelstein discusses the two versions, MELENCOLIA§Ia (figure 1.1) and MELENCOLIA§Ib (figure 1.3).

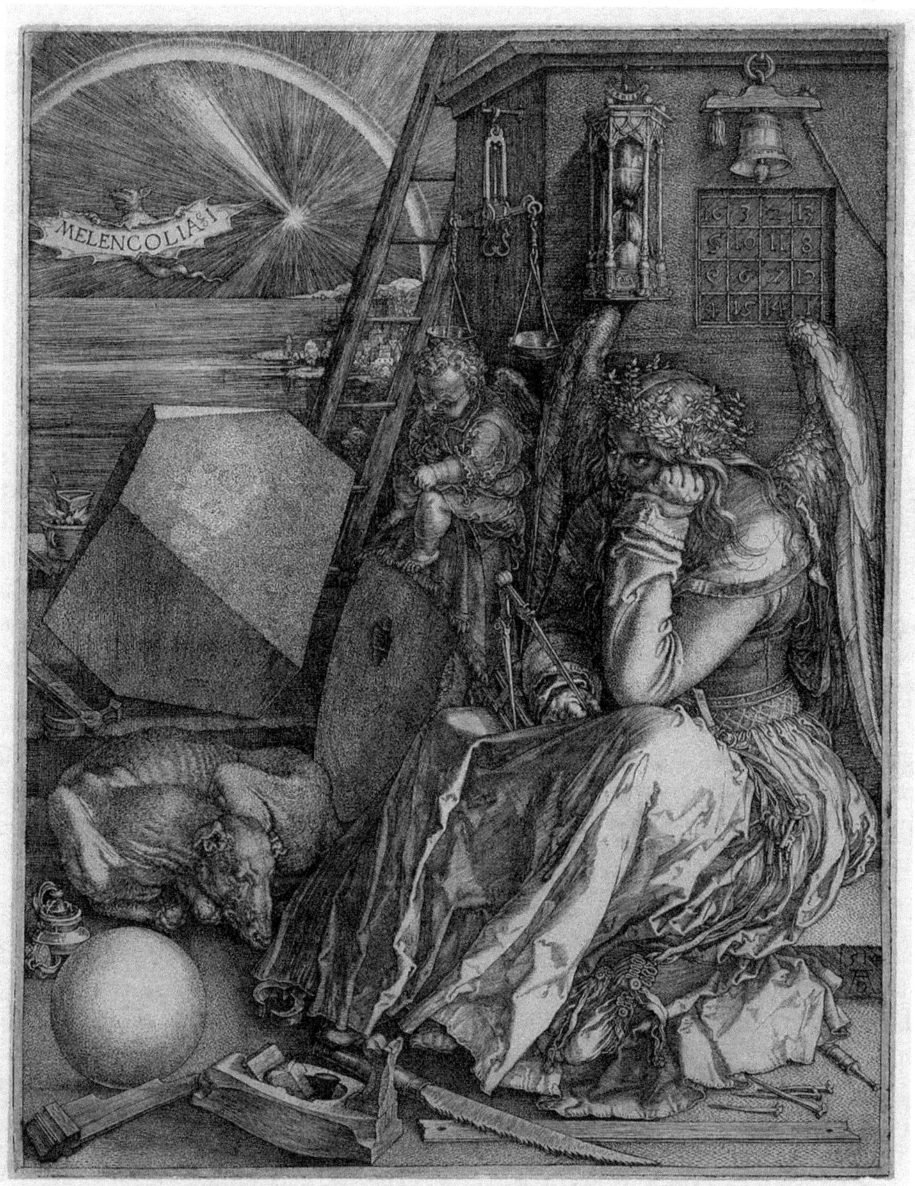

Figure 1.3. MELENCOLIA§Ib, Albrecht Dürer, 1514.

of applying for martyrdom. The unseeing knight may have represented Erasmus himself for Dürer.

In any case Dürer pays the knight a subtle tribute. Others have noted a subliminal face in *The Knight.*

- Part of the garment of Death forms a subliminal agonized face smashed between the knight's two fists.
- The knight may seem blind to Death but still fights him.
- A second subliminal face watches the knight from the rocks of the hillside.

Eventually, Dürer's friend Willibald Pirckheimer (1470–530) tells us, the high-minded Dürer would lament that the behavior of the Protestant clerics made even the Catholic clergy look respectable by comparison, and there is indication that Dürer never lost his high regard for Erasmus:

- Dürer's 1526 portrait *Erasmus of Rotterdam* has a gleeful subliminal face high on the left sleeve of Erasmus.

These examples indicate that Dürer's subliminal faces are not just puzzles and games. Some advance the story like a Greek chorus, telling us how Dürer felt about the subject of his art. For example, I suppose that the cliffs of Arco scowled because the humanist Dürer disapproved of what went on in the Arco cathedral. The *Young Couple* reminds us that soon after his marriage Dürer left his 15-year old bride Agnes in plague-stricken Nuremberg, and went to Italy and Pirckheimer. Then Death hiding behind the tree is the plague, and the bulbous-nosed beetle-browed figure that dominates the *Couple* could be the powerful Pirckheimer.

There are too many faces tucked away in MELENCOLIA§I for them all to have significant individual meaning. Likely they are there as further evidence of the unreliability of our perception. All our knowledge comes through the senses, to be sure, but we miss much and add our interpretation to what we do not miss.

3.4 The *Triumphal Arch*

Dürer's immediately preceding work, *The Triumphal Arch of Maximilian I,* is intensely programmatic and ideoglyphic, and so prepares us for MELENCOLIA§I. The code, the key, and the clear Latin version are all extant. It evolved in four stages:

1. Around the 5th century Horapollo Niliaci, an Egyptian grammarian from Phanebytis under Theodosius II (408–450 AD), claimed to render 189 hieroglyphs into plain everyday Egyptian [17]. In the 15th century, a Greek translation of his work, *Hieroglyphica,* was found.
2. In 1512/13 Willibald Pirckheimer translated *Hieroglyphica* from Greek into Latin. Dürer illustrated this translation and it was widely disseminated. It provided a fresh graphic vocabulary that influenced the visual arts of Europe.
3. Pirckheimer wrote a Latin eulogy of the Holy Roman Emperor Maximilian I, patron of scholars and artists and specifically of Dürer, who illustrated writings of Maximilian himself.

4. Finally Dürer translated Pirckheimer's eulogy word by word into dogs, goats, harpies, cocks, snakes, scepters, and other such glyphs, as the centerpiece of the *Triumphal Arch*, using the *Hieroglyphica* as a code book. For example, since *Hieroglyphica* states that the ancient Egyptians represented a king by a dog with an ermine stole, Dürer drew the good Emperor Maximilian as a dog so attired.

Centuries after Dürer, the Rosetta Stone discredited most of Horapollo's interpretations but not all. The clear text, key, and woodcut of the *Triumphal Arch* all survive. We do not have the clear text for MELENCOLIA§I but we might have all the keys.

Without the overall interpretation, it is easy to overlook significant elements in MELENCOLIA§I or take them for slips. Some suggest that Dürer could not spell his motto, and indeed its spelling is strange; that he could not draw a cube in perspective, since the octahedron is not quite a truncated cube; that he could not make clean corrections, since there are blurs and fragments in the magic square. All of these variations are likely to be intentional.

There are well-known limits to Dürer's knowledge, to be sure. As noted earlier, the astronomer Johannes Kepler pointed out that Dürer drew the ellipse that is produced when a plane cuts a cone as egg-shaped, with the end nearer the tip of the cone smaller than the other end [1, 18]. There is also a subtle error in optics in MELENCOLIA§I, which will be pointed out in due course. More to the point, Dürer morosely declared that there are ultimate limits to all human knowledge, including his own. He demonstrates them in the engraving itself.

Nevertheless, to catch his meanings we should begin by recognizing his unparalleled power with the burin, his mathematical skill in perspective, his dedication to precise language, and his acute observation of Nature. The peculiar spelling of MELENCOLIA§I, the odd shape of the octahedron, and the curious blur and other anomalies in the magic square, among other things, then become meaningful.

3.5 Mystery

Dürer could have explained his engraving for us. Leonardo made up many riddles and he wrote the solution after each, concealed only by his mirror-writing. Dürer and Pirckheimer left us the key to the *Triumphal Arch*; they could have done the same for MELENCOLIA§I. In his essay on this picture, furthermore, Camerarius ignored its many concealments, described the magic square as a spider-web full of dead flies, and omitted all its religious and scientific messages [16].[4] All this lack of candor calls for explanation.

I propose that Dürer and Camererius were evasive in their expression in order to evade the pyre. In their times religious thought-police persecuted both the old magic

[4] Editor's note: In an earlier draft Finkelstein writes, 'In 1514 the art critic had not yet been invented. The first published art criticism was the praise of this very engraving by a younger friend of Dürer, Joachim Camerarius (1500–74), who watched Dürer at work, translated his book on human proportions from German into Latin, and at the end wrote his eulogy, where he declared that although Dürer's art was great, it was the least of his accomplishments.'

and the new sciences with accusations of black magic and witchcraft. Catholic and Protestant church officials jailed or even burned people found guilty of the crime of 'innovation'. There was ample reason for Dürer and his heirs to mute their dissent.

Yet the market for this engraving soon wore out its copper plate. There may have been many eager and able to read its dangerous hidden message.

MELENCOLIA§I was not Dürer's first foray into dissent, moreover. In about 1498 Dürer published fourteen woodcuts of the *Apocalypse*. One of them, *The Opening of the Fifth and Sixth Seals*, shows the wicked hiding from God's vengeance. They are the Pope, Cardinal, Bishop, Kaiser, and Kaiserin, the top of the hierarchy of the Catholic Church and the Holy Roman Empire. Another, *The Battle of the Angels*, shows the imminent execution of the Pope by an avenging angel. So by the age of 27, 16 years before the date of MELENCOLIA§I, Dürer—with his master Wohlgemut—had already testified against the hierarchy of the Church and publicly killed the Pope and the Emperor in effigy. It seems that Dürer could express controversial views in his art. It suggests that the views lurking in this engraving are even more criminal, since he hid them so well that they remained unread by the general public for centuries. We can also infer that they would incense Protestant officials, since Dürer had already flaunted Catholic law unscathed. What was the crime of Albrecht Dürer? It is not visible on the surface of MELENCOLIA§I.

3.6 The demon

A black demon with the head of a mouse, wings of a bat, and tail of a serpent flies at us out of the engraving. It sports a banderole with the motto 'MELENCOLIA§I' that named the engraving. Since it carries a message, it is a messenger, literally an angel, but a dark one, to balance the light one on the other side of the engraving.

The motto presumably refers to the ancient Greek humor of melancholia, the black bile (*melan cholia*). It has been suggested that it also refers to, and even accepts, a revolutionary variation on the humor theory then being circulated by Agrippa. The evidence that Dürer had access to Agrippa is circumstantial but is accepted by Panofsky and Yates. Neoplatonic cosmology figures strongly in MELENCOLIA§I; a brief summary is called for. Today cosmology is revised drastically every century or so. The Neoplatonic cosmology permeated and dominated European thought for over a millennium. It described three 'worlds' or 'spheres':

1. World I is the terrestrial or elemental world, our material abode, made of the four elements distributed in space and time.
2. World II is the celestial world above, the site of astrological objects that govern the fates of empires, with one rigidly rotating crystal sphere holding the fixed stars and as many as necessary for the Sun, the Moon, and the planets. Ptolemy used over 90.
3. World III is the spiritual or intellectual world outside the celestial world, the home of angels and Platonic ideas.

Neoplatonists divided the human psyche into three corresponding 'mentalities' or 'faculties' particularly suited to these worlds [3]:

1. *Mens imaginatio*, the imaginative faculty, empowers the artisan, the artist, and the natural philosopher to view arrangements of the four elements in space and time, but nothing more.
2. *Mens ratio*, the rational faculty, allows astrologists and mathematical philosophers to predict when stars and empires rise and set.
3. *Mens contemplatrix*, the contemplative faculty, enables theological philosophers to discourse of angels and ideas.

The names for the three worlds vary. Agrippa also called them *idolus* or *imaginatio*, *ratio*, and *mens* [21]. The three worlds provide three philosophies, gateways to knowledge of God, and three books of wisdom:

1. Gateway I is natural philosophy. Book I is the Book of Nature.
2. Gateway II is mathematical philosophy. Book II is the Hebrew Bible.
3. Gateway III is theological philosophy. Book III is the Christian Gospels.

In Book I we see merely the creatures of God, not God, Agrippa wrote. We may use reason to deduce the existence and nature of God, but original sin clouds our souls and makes our conclusions unreliable. But Books II and III are the result of revelation and require only faith. Gateway III was the truest according to Agrippa. This would have made him no soulmate of Dürer.

Neoplatonism also kept the ancient Greek psychology of the four humors, one for each element: blood for air and the sanguinary temperament, yellow bile or choler for fire and the choleric temperament, phlegm for water and the phlegmatic temperament, and, last but least, black bile for earth and the melancholic acute depression. This ancient theory occupied Dürer's imagination in 1514 [2].[5] It survives today in our vocabulary for human temperaments.

Melancholia, the black bile, does not exist. It is a Greek myth invented to fit the mental illness of acute depression into the humor theory. Some said the appendix secreted it, some the spleen; neither secrete anything. Clinical depression exists, but not the black bile.

Astrologers had associated Saturn, the slowest, darkest planet, with the black humor, melancholia; with the lowest element, earth; and with the lowest people, earth workers. Agrippa was a revolutionary astrologer, out to elevate the downtrodden to the highest. He therefore elevated Saturn from the lowest planet to the highest, arguing that Saturn is highest from the Sun and most sedate. He attributed to Saturn the power to inspire all three faculties to their peak creativity, and transferred even the demi-godlike geometers—'earth measurers', after all—to the dominion of Saturn, among the farmers, miners, and ditch-diggers. He arrived at his conclusions by wordplay, not observation. For example, in his *Occult Philosophy* the touch of a diamond is said to demagnetize magnetite. Perhaps Agrippa deduced this by axiomatic reasoning from the name 'diamond', or adamant, whose root meaning is 'invincible'; certainly he did not infer it from experiment. Marlowe's Faust is said to have been based in part on Agrippa.

[5] Reference [2] is the standard work. It reproduces the Dürer works mentioned here.

Agrippa wandered Europe for much of his life, seeking royal support, making enemies, fleeing them, and leaving centers of intellectual ferment behind him [21]. He wrote a tract on the natural superiority of women over men, and defended a wealthy widow against witchcraft charges, and so, not long after this engraving, Martin Luther (1483–1546) linked Agrippa with the devil. In 1531 a plague broke out in Antwerp. The doctors left but Agrippa stayed to tend the sick. When the doctors returned, to recover their wealthy patients they filed charges against Agrippa for practicing medicine without a license. Agrippa fled once more.

In his later years he said that he had studied the occult philosophy because at one time he believed that it traced back to a secret divine revelation given by God at the same time as the public divine revelation of the tablets of the Torah. He remained in the Catholic Church and eventually adopted an uncritical fideism. His work on the uncertainty and vanity of the sciences exalted faith in Scriptures and divine revelation over all the arts and sciences, including astrology, and retracted his *Occult Philosophy*, but this had appeared in print earlier.

Panofsky and Yates saw many Saturnian melancholic elements in the engraving, in either the depressive sense of Ficino or the creative sense of Agrippa. They inferred that Dürer adopted Agrippa's esoteric philosophy, at least for the purpose of this engraving.

This does not do justice to Dürer as artist-scientist and follower of Leonardo. To be sure, MELENCOLIA§I uses the language of Neoplatonism, but there was no other. Nevertheless it is more radically humanistic and rationalistic than Agrippa or Luther. Agrippa, Erasmus, and Dürer all took the Gospel as divine revelation, but they took different paths through the Reformation. Agrippa's protest against the corruption of the monks who ran the Inquisition became radical [13] but he stayed within the Roman Church. Erasmus remained loyal to the Roman Church and distanced himself from Agrippa. Dürer left the Roman Church to follow Luther.

Dürer probably rejected astrology years before Agrippa, still an astrologist, came to Nuremberg. In 1514 Copernicus (1473–1543) had not yet published his heliocentric cosmology, but the mathematician and philosopher Cardinal Nicholas of Cusa (1401–1464) had long demolished the Ptolemaic spheres, and Leonardo had already written that the Earth moves, not the Sun, and that 'those who have chosen to worship men as gods—as Jove, Saturn, Mars and the like—have fallen into the gravest error'. Leonardo had derided necromancers and excluded astrology from his intensely rational works on astronomy. Similarly, a 1494 picture of the folly of 'Attention to the Stars', plausibly attributed to Dürer, is found in the *Ship of Fools* of Brant [7]. It shows the astrologer-fool immersed in a motley flock of misshapen fowl flying in random directions, perhaps representing the astrologer's quackery.

Martin Luther nailed his theses to the door in 1517 and Dürer soon became an ardent follower. Luther specifically linked Agrippa's *Occult Philosophy* with witchcraft. Dürer looked to the future for knowledge and Agrippa became the past.

Some say that the motto labels the angel opposite the demon. But in his other works Dürer puts his labels on what they label. Since the demon bears the melancholy label, it can reasonably be identified as Agrippa, who was Mr Melancholia of 1514. If Dürer accepted Agrippa's teachings, at least for this engraving, it seems unlikely that he would represent Agrippa as a demon. The *Hieroglyphica* gives a less damning meaning for a bat symbol, but the demon is not a bat. It is more likely that Dürer represented Agrippa as a demon in order to demonize him, for his astrology and for his adherence to the Church of Rome. Then the demon is not a messenger of the divine light in the heavens behind him, but a fugitive from it.

In this surface reading the '§I' in the motto is meaningless, except as a hint that there is something more.

3.7 The octahedron

Dürer seems to put his octahedron—the name given this solid by Federico [10]—in a place of honor. The artist stands before it and it balances the angel in the composition. Yet it is bleak and sterile against the angel. One looks for a deeper significance to justify it artistically, or at least to understand its shape. Most symbols in MELENCOLIA§I can be plausibly assigned to one of the three Neoplatonic worlds. Which is the world of the octahedron?

A rhombus is a four-sided planar figure whose sides are all equal. A rhomboid is a closed surface formed of six congruent rhombuses. All the descriptions I have encountered see the octahedron as a truncated rhomboid with its longest diagonal vertical[6] [25].

If the octahedron represents any of the Neoplatonic worlds, then its mathematical nature and its altitude in the picture unambiguously mark it as the intellectual World III, the abode of Platonic ideas and angels. The ghosts in the octahedron also support this conclusion, which puts them in philosophical Heaven. I adopt this interpretation in what follows.

If the two missing vertices of the octahedron were restored, they would seem to lie in a vertical line in space. The compulsion to see the figure as a symmetrically truncated rhomboid is strong. It then has a vertical axis of three-fold symmetry.

In principle, however, one cannot determine the form of a solid from a one-perspective view. One can shift any surface point along the line of sight from the eye without changing its apparent location in the perspective view. Therefore many different solids have the same perspective view. The 'reversing staircase' and the reversing Necker cube exhibit this ambiguity. The ambiguity of perspective is an important limit to our knowledge of the world. For example, a single view of the stars and planets gives us no inkling of how far they are.

Sometimes other cues are so strong and unconscious that we cannot help inferring a particular solid. For example, it is hard *not* to see a perspective of a cube as a cube, or a perspective of a person as a person, even though intellectually we know the picture is ambiguous. The Dürer octahedron is subtler than the illusions I have mentioned, in that they are merely ambivalent, while the Dürer octahedron is at least trivalent.

First, view the octahedron with head erect, and see the First Ghost on a truncated rhomboid. Second, view it with head cocked to the right, and the First Ghost,

[6] Mathworld, http://mathworld.wolfram.com/DuerersSolid.html, describes the octahedron thus.

rhomboid, and threefold symmetry axis disappear from our perception. Instead one sees the Second Ghost on a nearly rectangular slab with two diagonally opposite corners trimmed, cocked, and cantilevered back toward the horizon. Third, turn the engraving onto its left edge to see the Third Ghost, and the octahedron changes shape again. It changes its shape when we turn it.

Some claim to have measured the angles of the octahedron from the engraving. This is clearly impossible. These angles depend on implicit assumptions that are not deducible from the engraving. This impossibility is the point of the octahedron. We can never know its true shape from this one-perspective view.

If one compresses the engraving vertically by a factor of $\sqrt{\varphi}$, its frame becomes nearly square, and the octahedron could pass for a truncated cube; but only until one juxtaposes a perspective of a true cube for comparison.

One permissible interpretation sees a truncated rhomboid with the apical angle close to 80° [20]. If one juxtaposes a model of such a rhomboid with the engraving, the match is good [19]. Before the engraving Dürer made what seems to be a rough sketch of the octahedron that has survived, although it was recognized as such only recently [26]. The sketch shows an irregular pentagon inscribed in a circle. The apex angle in the drawing is 79.5° ± 0.5°, close enough to MacGillavray's interpretation.

The sketch also shows what seems to be a regular heptagon. An irregular pentagon is inscribed in the heptagon by omitting two vertices of the seven, separated by one vertex between them. There is no way to make a regular heptagon by Euclid's methods, but Dürer gives a simple approximation in his posthumous work on descriptive geometry. Theoretically, the sharpest angle of the inscribed pentagon is three fourteenths of a full circle, or about 77.1°. Perhaps Dürer merely copied his imprecise sketch, for the exact angle is unimportant for the illusion, as long as it is close to a right angle but not too close.

Leonardo had already undertaken a mathematical theory of beauty. That goal would occupy much of Dürer's later years, and result in his works on human proportions and on how to construct projections and perspectives with a compass and ruler. Dürer's study of human proportions is a dry gallery of stark outline drawings of standing human nudes, with tables of anatomical dimensions given to one part in a thousand. The human measurements were to be the foundation of a geometry of beauty, as stellar measurements were the foundations of Ptolemy's geometry of the heavens, and later Kepler's [18]. Music was already divine and part of the quadrivium because it was considered mathematical. By providing a mathematical basis for art, Dürer hoped to sanctify the graphic arts as well. Possibly this belief in a mathematical theory of art was only one aspect of a belief in a general mathematical wisdom.

I propose that Dürer designed the octahedron to be ambivalent, irresistibly construed as a truncated rhomboid in one orientation, as a truncated slab in another, and as something else from yet another. For this reason he had to both stretch the cube and truncate it; neither alone would have created an ambivalent solid, for we are ready to see cubes in any orientation. The exact angles of the octahedron are then immaterial as long as they are not too close to right angles, which might force themselves upon our perception in any orientation. The octahedron is a puzzle that is intentionally unsolvable. It declares that the intellectual world has a mathematical design, but that this design is inaccessible to us.

3.8 The angel

In heraldry, angel-wings mean sanctity, eagle-wings fame. The sacred wings of MELENCOLIA§I lightly touch the hourglass, the scales, the numeral 1 in the Dürer table, as if by accident, and nothing else in the picture. No one can accuse Dürer of declaring that these scientific tools are blessed by the angel, above the workman's tools that litter the ground. She holds a sealed book and an apparently idle pair of compasses. She is not holding the compasses as one who uses them would, nor has she a suitable working surface.

Some commentators say that she is the spirit Melancholia herself. The motto on the banderole, the shadow on the angel's face, and the fist on her cheek are indeed consistent with melancholia. But the hidden motto is not negative in mood but positive and in any case is attached to the demon, not the angel. Her wreath of water-cress and water-ranunculus protect her from earthy melancholia by their water nature [20], so she cannot be melancholy according to Agrippa himself. The angel's expression is alert and focused, not soft and sad. Above all her gaze is not downcast. Downcast gaze and downturned corners of the mouth may be innate physiological responses to melancholia, transcending culture and language. In any case they were mandatory for portraits of melancholia. The angel's gaze is elevated, not depressed, and her lips are not visibly downturned.

The engraving has two states, which I call MELENCOLIA§Ia and MELENCOLIA§Ib.[7] Comparing them reveals some of the engraver's intentions. In MELENCOLIA§Ia there is a small upward line at the near corner of her mouth that can be read as a smile or not. The far corner of her lips is almost hidden by the turn of her head, so that only a trace of smile, if any, is visible there (see figure 1.3). The whole meaning of the angel's expression, of the angel as a whole, and of the entire engraving, is controlled by an almost imperceptible trace of the burin, creating a high-intensity focal point for attention. This focusing device is found in other Dürer works. In MELENCOLIA§Ib, however, the angel has a darker face. This does not mean that she is not smiling, but leaves it for us to decide. MELENCOLIA§Ib is more serious in tone than MELENCOLIA§Ia, but also more ambiguous.

The ambivalent smile of the angel would seem to be Dürer's homage to the ambivalent smile of *La Gioconda*. The *Mona Lisa*'s smile is wry. Her lips curl upward significantly more on her left side than on her right, so that the viewer oscillates restlessly and irresistibly between two interpretations. This is sufficiently unusual in portraiture that when Dürer's angel too shows an ambiguous expression it is natural to consider the possibility of influence, and then to look for more. It is easy to find some. The bird's eye view of background watery landscape in the *Mona Lisa* was an innovation of Leonardo, demonstrating his mastery of perspective.

[7] Editor's note: Historically, MELENCOLIA§Ib is the first of the two engravings. But as this does not affect Finkelstein's analysis, they remain as he has labeled them.

Dürer adopts such a perspective for the background of MELENCOLIA§I. There is no direct evidence that Dürer ever saw the *Mona Lisa*, but neither is it excluded, and others have considered it likely that Dürer saw works of Leonardo. Since he had traveled for a month from Nuremberg across the Alps to Venice to study Italian perspective for over a year, it would have been unnatural not to cross the smaller mountain between Venice and Florence to see the great Leonardo, the master of perspective. The engraving supports this supposition with its distinctive facial expression and background.

The two expressions differ significantly, however. We do not doubt that La Gioconda smiles with at least half her lips. We see them clearly. Only what she feels puzzles us. But we can never see clearly whether the angel is smiling or not. If she moved her sleeve by a millimeter or turned her head by a degree the ambiguity would dissolve. Dürer posed her so that the angel's expression could be taken as serious or smiling depending on the expectation of the viewer. This is a picture of not a facial expression but ambiguity itself. The pose shows contemplation, as in Rodin's *Thinker*, not melancholy. This and her wings suggest that she is the contemplative faculty, proper to the intellectual world of Gateway III, theological philosophy. Let us see how this interpretation fits the rest of the engraving, and why Theological Philosophy is smiling, if she is.

3.9 The compasses

Whenever the quadrivia are personified in the 16th century, Geometria holds the compasses. Since the angel has compasses, some say that she is Geometria. In a Dürer woodcut of 1504, however, *The Astronomer* measures a globe with compasses under a full moon, so compasses also occur in the celestial world. Dürer's compasses do not identify their bearer definitively.

Some say the angel has measured the stone globe that lies before the dog. But the astronomer studies his globe intently while the angel looks right past this globe into space; as indeed she should if she is Contemplation or Theological Philosophy and the globe is the elemental world. Lines of sight matter, especially in the work of a projective geometer.

To use compasses one must control both of its points at once, either by holding the apex or grasping them both. The angel grasps only one arm of the compass and that near the point, and she has no drawing board or table beneath the point, so she cannot use them as compasses. If anything, she seems to be sticking herself in the thigh and smiling. Since this is absurd we must look deeper. A few centimeters will do.

3.10 The Arab

The angel's right knee is also the top of a hood worn by a man, shown from head to foot, the grandest subliminal element in the engraving. His profile is cocked at about 45° to the left and looks downward. He has a full black beard, an aquiline nose, Arab attire, and a black headband across his brow. Call him the Arab for the nonce. His face eventually leads us to his powerful shoulder, almost a hunchback, on which the angel rests her elbow and compass point. His left arm reaches toward the saw/sword on the ground. The left foot of the angel, where it extends beyond her robe, is also his left hand, and her toes his fingers. In the other direction his shoulder leads us to his torso, waist, and knees. The angel steps on the hilt with her right foot as though to foil him.

Now one can see what the angel is doing with her seemingly useless compasses. She is sticking the Arab. His right hand, which doubles as her spindle, reaches up past his face to fend off the compass point, expressing pain.

This reminds us that Dürer's parents may have been refugees from the army of Muhammad II, judging by when they left Ajtás. I am reluctant to suggest that Dürer's angel half-smiles as she surreptitiously pricks the Arab, but Dürer's knight too seems oblivious to Death and the Devil as he crushes one of them between his fists. The sneering cliffs around the Arco cathedral are acts of concealed aggression by Dürer himself. At least we can cast the angel as the contemplative faculty with more confidence, now that we understand her anomalous compasses.

We have already read the angel as the contemplative Faculty III of the theologians. Dürer seems to show that contemplation does not lead to one absolute truth but to too many quarreling faiths.

The serrated sword for which the Arab is reaching suggests that Dürer associated Islam with militancy. This is plausible. Hungary had been mauled by both Christian and Islamic combatants in the 15th century. The first response of Muhammad II to the Crusades was barely stopped by the forces of John Hunyadi, who became a Hungarian national hero thereby. Later the Ottoman Empire would include Buda and reach to Vienna for a time. Since the fall of Constantinople was so epochal for Christianity, Dürer might conceivably have had its conqueror Muhammad II in mind. But since Muhammad II was more warrior statesman than theologian, Dürer would associate him with the rational faculty of the celestial world, not the contemplative faculty of theological philosophy.

A more plausible original for the Arab is the Prophet Muhammad himself. He would be proper company for the contemplative faculty of theologians.

3.11 The millstone

The most inscrutable of the three stones in MELENCOLIA§I is the stone wheel on which the boy—the *putto*—perches. Panofsky and Yates call it a grindstone but Doorly calls it a millstone [4]. Its perimeter is broken, its face is smooth. It would make an oversize bone-rattling grindstone but a passable millstone so let us go with Doorly. This is also the only stone in the engraving that could serve Jacob first as pillow and then as pillar, if he could raise it.

Since the stone globe represents the elemental world of natural philosophy, Gateway I, and the stone octahedron for the intellectual world of theological philosophy, Gateway III, by elimination it would seem that the millstone should somehow stand for the celestial world, the realm of reason and mathematical philosophy, or astrology, Gateway II. This representation would have to be obvious to his intended viewers as well. It was not to me.

Question: How is the celestial sphere like a millstone?
Answer: They both turn slowly and rigidly about a nearly fixed axis.

In olden days there were few such objects, and the millstone is the most common. For centuries before Dürer, cultures as diverse as Babylon, Greece, Arabia, Scandinavia, and Rome represented the celestial sphere by a millstone. Later, Galileo would too: 'Next, applying this reflection about the millstone to the stellar sphere …' [12].

In the widespread myth, the celestial millstone wanders off its axis into the celestial sea, presumably the Milky Way. This is the slow precession of the equinoxes from one house of the zodiac to another [22].[8] Santillana and von Dechend named it Hamlet's millstone because it is mentioned in passing in the legend of Ambleth, the original Hamlet, as retold by Saxo Grammaticus in the late 13th century [23]. Saxo was first printed in Paris in the very year 1514 of MELENCOLIA§I, indicating that the trope of the celestial millstone was still alive in Dürer's day.

According to the myth, in the Golden Age the millstone ground out gold. That is, in the good old days the stars were propitious. In later, lesser times it ground out salt, making the ocean undrinkable. In our disastrous age it produces sand and the terrible maelstrom, whose very name invokes the myth. The fault is in our stars, not in ourselves. Saxo's Amleth, Shakespeare's Dane, and Dürer's boy share the themes of melancholy, parental ghosts, dynastic overthrow, and the millstone. This suggests that Dürer and Shakespeare both transmuted personal grief into the Agrippan creative melancholy that inspires great works, one mourning his mother Barbara, the other his son Hamnet, and that both infused their ensuing creation with Saxo's tale of familial bereavement; whether independently or not I cannot say [15].[9]

The faculties are out of their proper spheres. A celestial revolution is going on in this engraving. The seated angel, the contemplative faculty of the intellectual World III, the highest, is grounded and lowest. Dürer has demoted theological philosophy.

The boy, the imaginative faculty of the elemental World I, now perches on the celestial World II. Dürer has promoted natural philosophy above theological. The natural philosopher observes matter in space and time, traditionally the lowest world, but thereby approaches God. The scientific revolution is in the seating plan.

The boy watches passively and draws what he sees. We know that Dürer himself did experiments, since he invented drawing aids and etching, but he has not made them part of his philosophy or his boy. The pioneers of experiment, Galileo and Francis Bacon, have yet to be born.

The equality between the angel and the boy, therefore, is not the Hermetic equality between the worlds above and below, as has been suggested [3]. The Hermetic doctrine of the Faustian magus suggests a magical influence of the higher levels of a hierarchy on the lower. It clashes in spirit and letter with the rationalism, humanism, and humility of MELENCOLIA§I. When Dürer rolls away the celestial

[8] See [22] for argumentation for the astronomical interpretation of all myth.
[9] See [15] on Hamlet and Hamnet.

sphere, he eliminates the Hermetic above and below and unifies the heavens and the Earth, preparing the way for Newton.

Hermetic doctrine relates the elemental world to the celestial and becomes irrelevant when the celestial millstone is taken out of service. 'As above, so below' may well have been Agrippa's credo at one time but it is not Dürer's. Like Leonardo, Dürer caricatured the idea that events in the sky foretold events on Earth. He printed maps of the Earth as seen from the heavens and the celestial sphere as seen from the Earth. They are realistic maps of the stars above and the planet below and indicate no similarity or correspondence between them. The unity that Dürer symbolizes by rolling away the millstone does not hark back to Hermes Trismegistus but leaps ahead to Newton.

3.12 The dog

The dog was the problem from the start. MELENCOLIA§I, *The Knight,* and *Jerome* each have a dog and an hourglass. Their hourglasses remind us that life is brief. What do their dogs mean?

The dog of MELENCOLIA§I dozes rather too close to the straight line of the comet, octahedron, and globe to be an accident. He is just where the celestial world would lie in the old cosmology. Someone rolled the battered millstone away and leaned it up against the house and the dog lay down in its place. The geniuses of the celestial world were mathematical philosophers, astrologers, statesmen, kings; why dogs? To us, dogs represent love and fidelity, to Agrippa, they represent love or flattery, but to Dürer, somehow, they represented the rational faculty of Agrippa's World II. We may infer this by elimination and guesswork, but we also have to understand how Dürer's viewers could have understood this on sight.

To read this dog I first looked at Dürer's other dogs. He had drawn at least two the year before that are extant, in the *Triumphal Arch* and the *Hieroglyphica.* The dog drawn by Dürer for the *Hieroglyphica* is like the one in MELENCOLIA§I, and some take it for a sheep. Perhaps the breed no longer exists.

In his *Triumphal Arch* Dürer represented the Emperor Maximilian I as a dog with a stole, as the *Hieroglyphica* required. The *Hieroglyphica* also tells us that the Egyptians represent a prophet by a dog, 'because the dog looks intently beyond all other beasts upon the images of the gods, like a prophet' [17]. For whatever reason, this dog is a prophet, the prototypical worker of the celestial sphere, an astrologer.

Reading the language of *Hieroglyphica* is trickier than writing it. Horapollo wrote that a naked dog can also represent an embalmer. This is possibly a reference to Anubis, the jackal-headed god of embalming. And a dog can also represent the spleen, odor, laughter, sneezing, and rule.

Others have opted for the spleen, influenced by humoral psychology, but this does not fit the story as snugly. The main people associated with the celestial world and with inspired rational faculty, according to Ficino, are the prophets and the kings. The *Hieroglyphica* represents both by dogs, naked or stoled. So Dürer obediently represented them both by dogs on the *Triumphal Arch* and in MELENCOLIA§I. His dogs seem to be revitalizations of Egyptian clichés of dog- or jackal-headed deities.

The dog as Faculty II of the celestial world fits all three master engravings rather well. The dog of *Jerome* can sleep soundly because Jerome's melancholia inspires his Faculty III, theology, not his Faculty II, astrological prophecy. According to traditional Catholic apologetics St Jerome did not believe in astrology. The dog of MELENCOLIA§I rests with half-closed eyes because his celestial world is rejected. *The Knight*'s dog runs beside the knight because the knight is engaged in affairs of state, with inspired Faculty II. The knight might be Erasmus, who believed in astrology [8]. Now the three dogs increase one's confidence in the three-world interpretation of the master engravings.

3.13 The wave

Following my hermeneutic rule I first looked for Dürer's millstone in the Bible. Of 12 millstones in the Bible, all but one are metaphors for livelihood or industry and refer to no other element of the engraving, so put them aside. What remains is:

> And a mighty angel took up a stone like a great millstone, and cast it into the sea, saying, Thus with violence shall that great city Babylon be thrown down, and shall be found no more at all. (*Revelation* 18:21, King James)

The engraving and this one sentence of *Revelation* share the mighty angel, the millstone, the sea, and even melancholia, for *Revelation* imbeds this millstone in a score of verses lamenting the destruction to come. The millstone of *Revelation* is therefore a reasonable candidate for Dürer's millstone, and the only one in the Bible.

But now that we look at it, the Biblical millstone is as puzzling as Dürer's was at first. Babylon was the great empire of the ancient world. Why would dunking a millstone show how great a force would destroy Babylon? A more cosmic image—say, sinking a continent—would seem to convey power better. A more domestic image—say, breaking a wine-glass—would show how easy it would be for God to destroy Babylon. But a millstone is neither here nor there. Mere people can drop a millstone into the sea, and it does the millstone no harm. If we take the account literally the angel of *Revelation* is merely a bit peeved with Babylon.

On the other hand, if the millstone of *Revelation* too were the celestial world, throwing it into the sea would be a cosmic disaster, the very same one as Ambleth's millstone rolling into the sea, the wobble in the celestial sphere that ends one world age and begins another [22]. This would make a metaphor fit for an emperor.

Any biblical reference to the celestial millstone might well be veiled, because it is a vestige of the same Babylonian astrology that Abraham fled. The separation in *Genesis* of the waters below from the waters above by the celestial sphere seems to be another biblical trace of the three-world cosmology.

If this verse is intended by Dürer then one would expect to find Babylon and its downfall in the engraving too. Indeed, it has already been noted that the engraving subliminally shows a wave destroying a city by the sea with ships in port, without even calling in the *Revelation* [16, 24].

Over the leftmost pan of the scales, through a small triangle framed by the ladder, house, and pan, a closely spaced system of fine parallel wavy lines flows from behind the house leftward into a great wave that looms high over the city, about to crash down on it and wipe it out.

The wave too is ambiguous. It can also be merely the shoreline beyond the city, not looming over the city at all. Then the lines are merely geological strata, not water. We cannot tell whether they are near the city or far from our one-perspective view. The ladder hides exactly the connections that would force one interpretation or the other. If the shore had a smooth C-shaped curve we would tend to see it as horizontal, but the visible shoreline is oddly straight and vertical on the canvas, evoking our propensity to recognize vertical lines. Dürer has also shaded the wave darker than the remote hills, to allow us to separate them. He allows us to see this line either as a vertical wave or a horizontal shoreline. At this point I realized at last that this recurrent perspectival ambiguity was not a weakness of Dürer's art but the very point of the engraving, driven home again and again until even I could not miss it.

3.14 Destroying angels

In MELENCOLIA§Ia the ambiguous vertical shoreline-or-wave-front framed between the fourth and fifth rungs of the ladder can also be seen as a face of a man or woman. Then it is not hard to find four destroying angels in the wave. They are present but completely reworked in MELENCOLIA§Ib, so probably they are intentional. The most evident change is in the lip line of the leftmost destroying angel, whose profile is the shoreline or wave-front.

With a single slant stroke of the burin Dürer changed from a smile of *Schadenfreud* in MELENCOLIA§Ia, to a frown in MELENCOLIA§Ib. To their right swims a fish, endorsing the wave theory. All these were obliterated by Jan (Johannes) Wierix,[10] defender of the faith.

Now the engraving and the verse of *Revelation* share the millstone, angel, sea, Babylon, destruction, and melancholy. This also reinforces the connection we sketched in lightly between Dürer's millstone and Hamlet's. I assume that the millstones of *Revelation*, MELENCOLIA§I, and Ambleth, in different centuries and continents, all represent the celestial world, and that Dürer's millstone refers to *Revelation* as his comet refers to *Genesis*.

Revelation would have referred to the end of the Age of Aries and the beginning of Pisces. Dürer, however, would have referred to the end of Pisces and the beginning of Aquarius. Since Dürer had already executed the hierarchy of the Church *in absentio* in his *Apocalypse*, he might have meant the same by the destruction in MELENCOLIA§I. Then MELENCOLIA§I is in the future tense, as *Jerome* is in the past and *The Knight* in the present. This suggests how to hang the triptych. Put

[10] Editor's note: Finkelstein refers to Wierix's 'sanitized version' of Melencolia I from 1605. In an earlier draft of this manuscript Finkelstein writes, 'Jan Wierix (1549–1615), the Flemish Dürer, made from scratch (so to speak) a picture that is also called 'MELENCOLIA I' and is usually considered to be a copy of MELENCOLIA§I.' For a copy of Wierex's 'Melencolia I' see http://www.metmuseum.org/art/collection/search/391254. Other changes Wierix made to Dürer's image are discussed later.

Jerome on the right, the knight in the center, and the boy Dürer himself on the left. Read them from right to left.

3.15 The ladder

Any artist who put an endless ladder next to an angel, a house, and an upturned stone in 1514 could be sure that the educated viewer of the time would see Jacob's ladder, angel, house, and stone of *Genesis* 28. The upper end of the ladder is unseen, so it may indeed be in Heaven, but it is surely not the ladder of Dürer's crucifixion scenes. Ladders as gates of Heaven occurred commonly in earlier art, including Dürer's own. The most relevant verses are:

> And he dreamed, and behold a ladder set up on the Earth, and the top of it reached to Heaven: and behold the angels of God ascending and descending on it. *King James Bible, Genesis* 28:12
>
> And he was afraid, and said, How dreadful is this place! this is none other but the house of God, and this is the gate of Heaven. And Jacob rose up early in the morning, and took the stone that he had put for his pillows, and set it up for a pillar, and poured oil upon the top of it. *King James Bible, Genesis* 28:17–18

So the ladder is the gate of Heaven of *Genesis* and also of the coat-of-arms; it is another signature. *Beth El*, the house of God, is also the legendary site of the first house of worship of the God of Israel. Early on, therefore, *porta caeli*, the gate of Heaven in the *Vulgate*, became a metaphor for the Catholic Church itself. Since Dürer criticized the Church strongly, it is doubtful that he intended that metaphor here.

3.16 The house

The house too is polymorphic. From up close we see two blank walls supporting a miscellany of instruments. The side wall does not seem much wider than the boy leaning against it, about right for an outhouse or a chimney, not a house. Its front wall extends out of the scene to the right and holds a bell, a square array of numbers, an hourglass, and a sundial. Its left edge bisects the picture to within a millimeter or two. Its side wall holds a chemical balance or scales. The bell-rope trails off the edge to our right. Step back several meters or squint and the picture changes.

The square and bell together become a lattice window, the hourglass a bay window, the scales a side window, and the outhouse a full-sized house showing us three of its windows. The angel blesses all three windows with her wings.

Assembling familiar objects from unrelated parts in this way, for example, faces from fruits, or demons from vermin, was an optical illusion often practiced by artists of the era. Here there is allusion as well. If the house is what the sanctified instruments on its walls open into, it is the humanly perceptible universe, the physical cosmos, Agrippa's Book I. The ladder and the angel have already told us that this is also the house of God, and the instrumental windows tell us that music, measurement, and arithmetic look into it. Leonardo wrote that knowledge comes from experience, Agrippa wrote that scientific experience leads to God, and Dürer draws both with his burin. This juxtaposition of science and religion is inconsistent with Church teaching of the time, which provides one and only one gateway to absolute truth, but it is consistent with Dürer's known humanism.

3.17 The magic square

The matrix of numbers set into the masonry wall in MELENCOLIA§I like a window lattice is a magic square, meaning that it is filled with consecutive numerals starting from 1 and every row, every column, and both main diagonals add up to the same number. In a 4 × 4 magic square that number must be one quarter the sum of the integers from 1 to 16, or 34.

This is a gnomon magic square as well. This means that besides being magic, its four quadrants, its four corners, and its central tetrad add up to the same number. A gnomon is a pointer or indicator, in particular of a sun-dial. A carpenter's square was also called a gnomon, perhaps because it looks like a crude picture of a pointing hand. This is called a gnomon magic square, I imagine, because removing any of its quadrants leaves a figure that resembles a carpenter's square. In addition, the sum of any pair of numbers symmetric about the center of this square is 17.

In his *Occult Philosophy* Agrippa assigns a magic square—then called simply a table—to each of the seven 'planets' then known, in the ancient order of Saturn, Jupiter, Mars, Sun, Venus, Mercury, and Moon, the order of their apparent periods about the Earth. He gave each table in both Arabic and Hebrew numerals. These magic squares have 9th century Arab sources [2] but may be older.

Agrippa omitted the one-by-one magic square [5]—reserving it for God?—and there is no two-by-two magic square, so the table of Saturn is three-by-three, the table of Jupiter is four-by-four, and so on to the nine-by-nine table of the Moon. The number of Jupiter is then $1 + 2 + \cdots + 16 = 136$. The number of the Sun is the $1 + 2 + \cdots + 36 = 666$ of biblical fame, suggesting that these tables are older than the biblical passage about the Number of the Beast.

Gematria is a Hebrew numerology based on the fact that any letter of the Hebrew alphabet is also a number and that therefore every Hebrew word has a number, the sum of its letters. Agrippa used gematria to find reassuring Hebrew words in Jupiter's table and so prove its virtue.

As a child I wondered why such mundane arithmetical arrays were called magic. It is because they were sold and used as magical talismans. For example, Agrippa warned that an unshielded Saturn caused acute melancholia, and prescribed wearing Jupiter's table as a shield against the melancholy influence of Saturn.

Some say Dürer borrowed Jupiter's table from Agrippa [2, 3]. Perhaps they did not read the tables in question. Jupiter's table and Dürer's table (figure 1.4) are significantly different: Dürer performed permutations on Jupiter's table that left it both magic and gnomon in the arithmetical sense. He inverted it and he

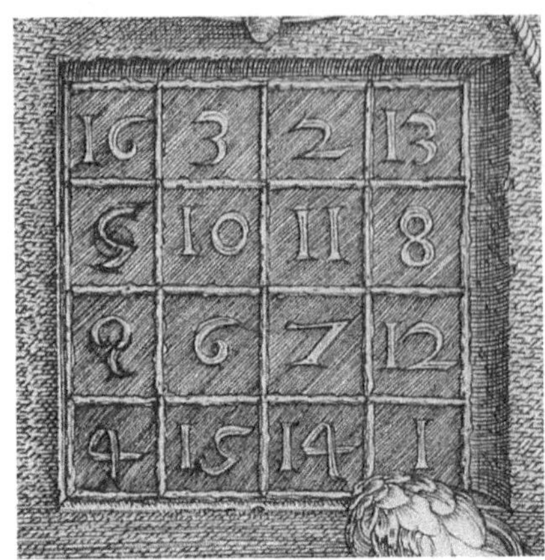

16	3	2	13
5	10	11	8
9	6	7	12
4	15	14	1

4	14	15	1
9	7	6	12
5	11	10	8
16	2	3	13

Figure 1.4. Top: Dürer's table (MELENCOLIA§Ia). Bottom: Jupiter's table (right) and Dürer's table (left).

interchanged the two middle columns. What did he mean by this, if anything? Such changes by the artist are both puzzles and clues.

The Bible, Agrippa, and even Horapollo fail us here, but there is a plausible reading. Dürer's table, unlike Agrippa's, echoes the dateline below it. Its bottom line is 4 15 14 1, flanking the date 1514 of the engraving by Dürer's initials in Latin gematria, where $A = 1$, $D = 4$.

It tells us something about Dürer's mind that he saw this possibility in the Jupiter table and then made use of it. And it transports us to an era before the Newtonian concept of a mathematical theory of Nature. I diffidently suggest that for Dürer this kind of rhyme between the table above and the date-line below was an intimation of an accord between divine mathematics above and secular reality below, whose discovery was a goal for the natural philosopher. The examples of a mathematical theory that he could build on, like the Pythagorean musical scale, the Archimedean law of flotation, Euclidean geometry, the crystal spheres of Ptolemaic astronomy, and his own descriptive geometry, aimed at representation, not prediction. Galilean and Newtonian ideas of a law of motion waited in the next century.

The five in the magic square is puzzling. There seem to be two figures superimposed. Some say that Dürer made an error and corrected it badly. It is dangerous to doubt the precision of Dürer's burin, and there is plenty of evidence that he could erase perfectly. It is safer to assume that the blur means something; but what?

Such image-duplication was used at the time to convey motion. The horse of *The Knight* has two hooves on one foot; the spinner in the Velasquez painting *Las Hilanderas* (*The Spinners*, circa 1657) has six fingers on one hand. Hand and hoof are the most rapidly moving elements of their pictures. Xeno taught that a thing in motion cannot have a definite position. It seems that Dürer and Velasquez believed him, and show motion by putting the moving object in two positions.

The foremost figure is a better five if we turn it upside down. Upright it is closer to an S but has a minute but clear serpent's head, jaws gaping and forked tongue protruding. And what we see of the background figure is more like a flourish *§* than anything else in Dürer's alphabet.

Other characters in the magic square are significantly distorted. Both sixes in the magic square are actually the medial form of the lower case s, a descendant of the Greek σ still in use in Dürer's time, and they too have serpent heads. The numeral 0 is the Uroboros, with its jaws at the 6-o'clock position. These Ss and serpents will be read in sections 3.27–3.28.

The two states of MELENCOLIA*§*I differ in many ways, most conspicuously in the fine details of the Dürer table. In MELENCOLIA*§*Ia, displayed on the internet (for example) by the Fine Arts Museums of San Francisco, the numeral 9 in the magic square differs from Dürer's usual nine and ours by a serpentine wriggle in its tail. In MELENCOLIA*§*Ib, however, shown on the internet by the British Museum, the nine is not only wriggly but reversed, even closer to S than 9.

3.18 The door

Then there is the curious matter of the door to the house. The house that we see as we step back has several windows but nowhere do we see a door. When the door-man draws the house of God without a door, he is likely making a statement.

It seems to repeat the major melancholy statement of the piece. The house of God has no doors, only windows, we can look in but we cannot enter. Absolute truth is inaccessible to humanity. The ladder is for the angels.

The inaccessibility of absolute truth was a common idea by Dürer's time. It was taught by Nicholas of Cusa and Erasmus, already cited, and would reappear in the next century in the writing of Vico. Dürer could expect viewers to make out this message at first sight. Before Dürer did this engraving he had already written that the human mind cannot know absolute beauty. In this picture he seems to be saying the same thing in several ways. We must not assume that he separated truth and beauty as cleanly as some claim to do today, in particular since he expected a mathematical theory of both. If science is supposed to be a mirror of nature then Dürer's rabbit and Audubon's birds count as great science as well as great art.

3.19 The boy

The boy's writing instrument has a crossbar at its top, and so is not a stylus or a piece of chalk as some say, but the burin again. His block is therefore not a slate but a copper sheet or possibly a block of wood. He is engraving, not drawing. For he is Dürer.

The boy's eyes are disturbing under magnification. I cannot see them as both looking in the same direction. His left eye could be looking toward the globe, while in MELENCOLIA§Ia a ring of white in his right eye gives the strong impression that the boy looks directly at us with that eye wide open. MELENCOLIA§Ib changes the ring of white into a more naturalistic crescent of white, but the boy is still looking at me with his right eye and downward with his left.

The boy then carries two distinguishing features of Dürer, his burin and his amblyopia. Dürer drew his mother wall-eyed, like the Third Ghost, but the boy is cross-eyed.

As Panofsky said, Dürer meant this work to show many aspects of himself. As Yates said, the boy is Dürer at work in the creative frenzy that melancholia can inspire in the divinely gifted artist [3]. But his burin and his eyes indicate that he is not a generic starving artist scribbling meaninglessly on a slate after all [2].

The boy is the central figure in several senses. The geometric central axis of the engraving is the vertical defined by the leftmost edge of the house. It passes almost exactly through the nearest eye of the boy, a common way to indicate centrality in Renaissance art. The line of the comet suitably extended also strikes that eye.

Yates suggested that the boy Dürer is engraving MELENCOLIA§I itself in a self-referential way. The boy's copper plate seems too small for that suggestion to be taken literally, and it does not fit the three-world cosmology. In the tripartite psychology of the times, the artist's special faculty is the imaginative one, that of the natural philosopher, so the boy represents natural philosophy and the imaginative faculty too, as the angel represents theological philosophy. Then he should be looking at the globe, the mundane sphere of the four elements in space and time, as Gateway I requires. And so he is; at least with one eye.

So it seems that Dürer has perched his rump on the celestial sphere. I cannot read this as modesty, or as a sign of respect for astrology. It rather reinforces the message of the anagram, recommending Gateway I over Gateways II or III.

3.20 The scales

On the side wall of the house hang a pair of chemist's scales. Extended, their lines pass through the vanishing point, so the scales are balanced. Perspective causes the line of the pans on the paper to slant slightly upward toward the vanishing point, and the line of the suspension slightly downward. One pan brushes the boy, the other the angel. I think we can say that Dürer has established a balance between natural philosophy and theological philosophy, unlike Agrippa's fideism. The line of the eyes of the boy and angel also goes through the vanishing point. They have exactly the same altitude, that is, holiness.

3.21 The globe

Chance governs the elemental world, mathematical law governs the celestial world, and divine law governs the intellectual world. Dürer's goddesses Fortuna and Nemesis both stand on globes, liable to roll in any direction at any impulse, because luck and fate are unpredictable [2]. Therefore Dürer chooses a globe to represent the elemental world, where chance rules.

It is indeed the lowest of the three stone objects in this picture, as it should be if they represent the three worlds. Dürer, like Nicholas of Cusa and Leonardo, replaced the naive tri-spherical geocentric Neoplatonist cosmology by a tripartite philosophy with three faculties and worlds that are no longer geocentric, concentric, or even places at all. And here he has retired the celestial world.

3.22 The moonbow

I have never seen a moonbow but again NASA provides the picture and the Bible provides the caption.[11] The only bow in the Bible sky is the 'bow in the clouds' that Noah saw after the flood, a supreme example of divine revelation, direct communication with divinity. *Genesis* does not say whether Noah saw his bow by day or by night. Dürer had no choice once he had set the scene in the dark of night. In *Genesis*, the light in the heavens and the bow in the clouds represent two of the great gifts of God, so the picture should be joyous, and yet it is dark and ambiguous again.

The viewer of any moonbow or rainbow lies on the normal to the plane of the bow that passes through the center of the bow. Dürer's watching eye should therefore lie directly before the horizon point beneath the highest point of the Moonbow. This is not exactly where he is located by the vanishing point of the building lines, which is under the eye of the demon, but it is only about a centimeter off.

When we look directly at a bow in the clouds, the light source must lie directly behind us, and very likely Dürer knew this from personal experience. The artist is standing before the demon, a bit to the left of the center of the bow. Therefore to make such a bow the Moon should be behind him and slightly to his left. But according to the shadows on the house and the face of the angel the Moon is behind the artist and well to his right. This optical inconsistency seems to have no meaning,

[11] See http://meteoros.de/halo/halo1.htm for an on-line gallery of portrayals of heavenly phenomena such as moonbows and moon halos, including several from Dürer's time, and http://antwrp.gsfc.nasa.govapo-dap010704 for a photograph of a real moonbow.

but we need not try to believe that Dürer made a mistake in his depiction of Nature. Had he followed natural law the bow would have been far out of the picture to our left. He wanted shadow on the angel's profile to leave open the possibility of melancholia, and God's promise in the sky to express faith and inspiration, a physically difficult or impossible combination in any one view. I suspect that he deliberately sacrificed optical truth for a greater artistic one, knowing that few would notice or care.

3.23 The comet

Though it lacks the curved tail of the most traditional renderings the bright light in the sky is a comet, not a nova. The NASA online collection of astronomical photographs shows comets quite like it. A comet is straight when it heads for the Sun and sunlight pushes its tail out behind it.

The great comet 1471Y1 was first seen on Christmas Day in Dürer's birth year, and Dürer wrote of seeing a comet himself in 1503 [9, 19]. The physical natures of meteors and comets were not yet known in 1514. Leonardo believed, and Galileo would still believe in the next century, and our very words still reflect their belief, that meteors were meteorological, weather of the high air; and comets too.

Since the Bible has already explained other elements of the picture, let us search it for this one too. The answer is swift and unique. The terms 'comet', 'shooting star', and 'falling star' do not appear. There are three cases of a star that falls in *Revelation*, but they do not fit the picture well. The only blinding light in the biblical sky is the original light of the divine creation. Then the starless sky would recall the time in the Bible story between the creations of light and the stars. The comet would then represent divine creation and divine revelation.

The *Hieroglyphica* has no comet, but it has a star, another glyph for God. I see no way to decide whether Dürer had *Revelations* in mind or the *Hieroglyphica* when he drew the comet/star, but the two interpretations are consistent, even mutually supportive. The comet tail points quite accurately at the boy's head. This suggests that God is singling him out for illumination, above the representatives of the other two gates to Heaven; and presumably his viewers would see it as an emblem of the Reformation doctrine that divine illumination requires no priestly intermediary.

The Creation is a joyous occasion: Happy Birthday Universe! The news is good but the tone is somber. The tension grows. And so does the ambiguity. This comet is as realistic as NASA's. Nothing in the picture compels us to see it as a symbol.

Even with all these clues, I could not figure out whether Dürer's gateway to Heaven was mathematical or religious in nature. This is a good example of the kind of projection to avoid in such a study. First of all, Dürer does not separate science and faith. On the right they entwine in the building and ladder complex. On the left, there is an unnaturally straight line from the comet or the light of God through the intellectual world of mathematical philosophy to the heart of the elemental world of natural philosophy.

The celestial world is not on that line, however. It is old and damaged and out of service. The dog remains, idle, dozing. Dürer graphically reduced the three worlds to two in this engraving. As Leonardo had explicitly said, the terrestrial and celestial worlds are run by the same laws, and are actually one. The Hermetic doctrine was 'As above, so below', but Dürer—and possibly humanist philosophy—dissolved the absolute separation between above and below. The new astronomy saw the Earth and stars as made of similar stuff governed by similar laws.

3.24 The hexagram

Dürer was a descriptive geometer, practiced in constructing views of bodies from all sides and in reconstructing a body from its views. I therefore looked at the octahedron from all sides, searching for his meaning. He likely did the same, for he invented the representation of such polyhedra as nets of polygons [19].

The ground plan of the octahedron viewed as a truncated rhomboid from a certain viewpoint is a Shield of David framed in a hexagon (see figure 1.5). The top and bottom triangles of the octahedron project into the two crossed triangles of the hexagram. Perhaps Dürer truncated the cube, rather than some other regular solid, in order to create this hexagram.

It would be anachronistic, however, to call this hexagram a Jewish star and to infer any philo-Semitism. Centuries earlier the Khazars of the Crimea had adopted the hexagram for their flag when they adopted Judaism, but in 1514 Nuremberg the hexagram was still mainly a magical device, an amulet and talisman, possibly referring to the Hebrews of the Bible but not to contemporary Jews.[12] The Shield of David was called that because it was supposed to shield its bearer from evil spirits, and it was more often found in churches than synagogues. A two-footed hexagram like Dürer's ground plan is seen on a German altar of Dürer's time. Some said the Seal of Solomon was the hexagram, and others the pentagram. Traditions firmed later that gave Solomon only five points to David's six, and stood Solomon's star on two points but David's on one. Not long after Dürer's death the Jewish community

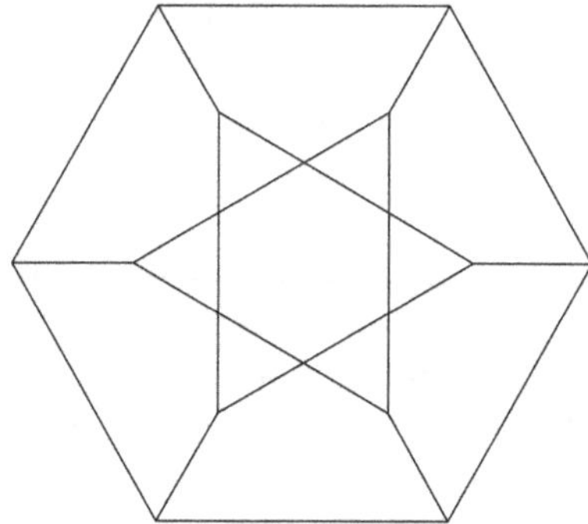

Figure 1.5. The hexagram in MELENCOLIA§I.

[12] See [14] for the concept of Hebraicism.

of Prague was granted the privilege of a flag and chose the hexagram, already associated with the Khazars, and this continued its evolution from magic charm to Judaic symbol. Dürer's use was likely a Hebraism, not philo-Semitism.

Hebraisms are conspicuous in the works of Dürer and Agrippa. Humanists like Dürer admired Jerome above the other founding fathers for Jerome's greater familiarity with the Hebrew Bible [2]. In his several drawings of Jerome studying Latin, Greek, and Hebrew Bibles, Dürer invariably placed the Hebrew Bible above the others, and higher is holier. His stereotype of First Worship[1] corroborates that the hexagram beneath the circumcised rhomboid was not philo-Semitism but Hebraicism. It connects the intellectual sphere, the Heaven of humanism, represented by the resulting octahedron, with the Hebrew Bible, as Agrippa did.

3.25 The bell-ringer

Dürer uses the vertical dimension on his plate in a traditional way: higher is holier. Jacob's angels get to Heaven by ladder. The angel and boy have equal divinity because they have equal altitude. The tools and nails of artisans litter the ground because they are mundane, while the instruments of science and mathematics hang nearer to the sky because they are holier. The horizontal dimension is cut in half by the leftmost wall of the house. Because the viewpoint is on the far left, it is not obvious at first that all three worlds of the cosmos and all four actors in the drama are to the left of the dividing wall, and only the house of God and the invisible bell-ringer are to its right. The point of view of the artist on the extreme left masks this clean cut by distributing elements of both halves of space on both halves of the paper, but all the biblical and intellectual elements are on the right-hand side of the median wall, and all the tools and nails are on the left-hand side. This seems meaningful and intentional.

The weighing scales are ambiguously located, hanging on the dividing wall itself. This detail fits our interpretation. Space and time were immaterial and therefore considered spiritual. Mass—'body' is the root meaning—was material, not geometrical and therefore not spiritual; it was not defined in terms of space and time until four centuries later, by Albert Einstein. For Dürer mass belonged to a lower world than time, and probably to the natural philosopher, not the theologian. I suggest that this is why the scales are to the left of the house of God while the hourglass and bell are to the right.

The crucible is no exception. If it were an instrument of alchemical science, as I thought at first, it would be elevated, but it is on the stone floor, so it is merely a tool of the goldsmith. Only perspective puts it high on the paper, above all other tools; perspective and filial feelings. This is further indication that Doorly read this correctly. This crucible is not only an alchemical instrument but also a remembrance of Dürer's goldsmith father, who died in 1502. Its position to the far left gives us one point on the time axis. The house of God puts biblical times on the right.

The octahedron is as opaque as stone. Dürer shows it hiding the divine light of Heaven from the inhabitants of the elemental world, the globe. The opacity of the octahedron in the left-hand side of the picture is the counterpart to the lack of a door in the house on the right-hand side. Neither mathematical nor theological philosophy can bring us knowledge of absolute beauty/truth.

The division of the picture by time agrees with its division by gateways, since Dürer took the occult philosophy represented by the demon on the left to be later than the theological philosophy and scripture represented by the house on the right. This probably fits his personal experience of them.

The near eye of the boy lies on the center line of the paper, making him the center of attention, but he too is left of the median plane. Dürer the artist-scientist takes Agrippa's Gateway I to Heaven, not his Gateways II and III.

Perspective provides a third dimension of depth into the scene along the line of sight. Dürer seems to use this expressively too. On the right side our view ends in the house and on the left there is an infinite vista. It seems that perspective depth in the picture is the familiar metaphor for depth of vision.

All the mortal creatures in the engraving are on the left-hand side of the median plane. The only one on the right-hand side is the bell-ringer, off stage, invisible. There can be little doubt about who is patiently, patiently holding the bell-rope, conducting the harmony of the spheres, and not to be depicted. Perhaps the engraving was intended for people who would prefer this invisible God to the pagan representation by Michelangelo in the Sistine Chapel, and perhaps find it more awesome.

3.26 Melancholy

The light and the bow and even the destruction of the city of evil are good news, cause for joy, and yet the picture is set in the gloom of night. Dürer has even transformed the rainbow of the common understanding into a moonbow. The comet illuminates nothing. Only the Moon behind us lights the scene and creates the shadows and the Moonbow. What might have been a sunny prophesy for Leonardo is dark as night for Albrecht Dürer. While Leonardo and Dürer were both mathematically ambitious, they differed importantly in their outlooks. Leonardo wrote confidently of 'a complete knowledge of all the parts, which, when combined, compose the totality of the thing which ought to be loved'. Leonardo was the rational optimist. Dürer on the contrary said shortly before this engraving, 'But what absolute beauty is I know not. Nobody knows it but God'. The darkness and the multiple ambiguities suggest that after all Dürer did not expect humanity to attain the *mathesis universalis* that optimists like Descartes and Leibniz had expected, and that Ramon Llull even claimed to have attained; and that he felt this as a great bereavement. He expressed this pessimism in the engraving atmospherically, by omitting the door of the house of God, and by numerous skillful demonstrations of ambivalence and relativity.

The most unambiguous aspect of this engraving is its ambiguity. Dürer tells us clearly that we cannot know anything clearly. The octahedron is a truncated rhomboid with vertical axis, or a slab tilted toward the horizon. There are four subliminal faces in the octahedron, or none. The city is to be destroyed, or not. The angel is sad, or smiling; the demon is good or evil. The house has windows, or not. The melancholy is sadness inspired by the inaccessibility of the absolute, or creative frenzy inspired by the influence of Saturn, or awe because the celestial world is off its axis. What we see depends on us as well as where we look. The elements of this engraving that I have discussed here are deliberately ambiguous in meaning or form or both.

What can we believe if we cannot believe our eyes? This questioning occurs famously in the first *Meditation* of Descartes (1596–650), who cites optical illusions as a reason for his method of universal doubt. MELENCOLIA§I anticipates Cartesian doubt, and Dürer's response to this doubt has a mathematical element that anticipates the *mathesis universalis* of Descartes and Leibniz, but already corrects it by accepting the limitations to human knowledge.

The ideal of a complete mathematical theory of beauty lies on the same long line of distinguished fantasies of mathematical wisdom as the number mysticism of Pythagoras and Plato, the *Ars Magna* of Ramon Llull (whom Agrippa studied) and Giordano Bruno (who studied Llull and Agrippa), and the *Ars Combinatorix* of Leibniz. Dürer does not deny the existence of absolute beauty/truth but despairs of knowing it. The boy withdraws from both theology and astrology and observes Nature.

Years after 1514, Dürer like Agrippa explicitly abandoned the search for absolute truth and beauty as futile and hopeless, crying out, 'The lie is in our understanding, and darkness is so firmly entrenched in our mind that even our groping will fail' [2].

In 1517 Dürer was still optimistic enough to join with Luther, but the engraving already expresses so much despair and darkness in 1514 that at first I doubted the dating of the engraving. In fact the date 1514 for the engraving seems reliable, since several studies for this engraving also bear that date. Rather, I infer that Dürer expressed his pessimism about attaining absolute truth in his engraving some years before he put it into writing. Perhaps because he was who he was, he drew first and wrote later. There is no need to doubt the dating of the engraving.

The night of the engraving is our benighted ignorance. Its darkness is the darkness 'firmly entrenched in our mind'. The mystery of the octahedron, crafted to be insoluble, is one of many metaphors in the engraving for our inability to see the absolute truth from our limited perspective. His chosen gateway to heaven is natural philosophy, in spite of its limitations. In acknowledging the limits of human knowledge, however, Dürer took a step toward modernity beyond the more optimistic Leonardo.

The posthumous publications of Dürer on measurement and on human proportions show that he never quit working toward a mathematical theory of beauty and truth based on measurement rather than abstract speculation, but his concept became more relativistic. For example, he included a variable horizontal scale-factor in order to represent human beauty of the lean, average, or plump variety as the reference frame of the artist requires. Agrippa inverted the primary meaning of melancholia just as Horapollo inverted the social status of the dog, the bat, and the serpent, making the lowest the highest. By using the ephemeral and counterintuitive languages of Agrippa and Horapollo, Dürer defies our ordinary rules of interpretation, making the work surreally ambiguous but also contributing to its remarkable endurance. This melancholia does not lead to death but to its transcendent opposite, a form of immortality, exhibited by the engraving itself, vibrant after five centuries. Actually no one in the engraving is merely sad. The overall darkness of the engraving expresses Agrippan melancholia, not depression but a creative frenzy, perhaps not just of the boy but also of the bell-ringer Pantocrator.

Dürer records that when he made a present of MELENCOLIA§I he sometimes accompanied it with *Jerome*, but never with *The Knight*. This pairing makes sense if the recipients were Protestants, who, for obvious reasons, could venerate Jerome and Dürer but not Erasmus. Jerome had criticized clerical corruption to the point of endangering himself, was a celebrated Hebraist, and was a close friend of Origen of Antioch, who is said to have held the universalist doctrine that ultimately everyone is saved, anathema to the Roman Church. Erasmus, on the other hand, remained loyal to the post-Nicean Roman Church and believed in the value of astrology as a hermeneutic art [8]. This theory requires only that Dürer gave the prints in question mostly to like-minded people.

3.27 §

A stylized S that I render as § separates the two words 'MELENCOLIA' and 'I' of the banderole. The same § stands before *The Knight*'s dateline and follows the date of Dürer's 1504 *Adam and Eve* in reverse. Likely all three have the same meaning. Perhaps it is reversed in one case so that it always faces what it modifies.

The numbers in the magic square are determined by its 4 × 4 size and the date line 1514. Once the bottom line is chosen the rest is no longer under Dürer's control and so probably carries no information. However, Dürer still controls the orthography of the numerals. Actually there are eight orthographically deviant numerals in the table of each version by Dürer, and only one in the Wierix version. Seeing them is easy now because they call attention to one another:

- The one that stands by itself is brushed by the angel's wing, and enlarged and elevated relative to all the other ones, as though blessed by the angel.
- The fours are not only numerals but also long solid bodies folded into crosses.
- Both fives are inverted.[13] Dürer ordinarily makes his five as we do today.
- One five has circular striations and a faint image of a numeral in its background.
- The two sixes are more open and overhanging than Dürer's usual sixes, more like σ than 6.
- The nine is open and curly in MELENCOLIA§Ia. Its lower end is sinuous, with two bends more than the usual nine. Dürer ordinarily made his nine as we do today.

[13] This was kindly pointed out to me by Bill Branch in a private communication. Branch also suggested the Ourobourous in connection with MELENCOLIA§I.

- The nine in MELENCOLIA§Ib is reversed,[13] more like a ς than a 9, and is less sinuous, but still has a serpentine reverse curve in its tail.
- The zero in MELENCOLIA§Ib is a serpent with its own tail in its jaws. The zero in MELENCOLIA§Ia has less detail.

Dürer's precision suggests that these small variations are significant and require interpretation.

When an S appears before a date in Dürer's work, it stands for *Salus*, or salvation [2]. Dürer prefixed the dates of *The Knight* and of the 1504 *Adam and Eve* with an S as if to say 'In the year of God 1514' or '1504'. The S also occurs in reversed form as the flourish following the date in *Adam and Eve.* I suggest that the § in MELENCOLIA§I is the same symbol of sanctity.

The Greek alphabet has two forms of our S, the medial (and initial) form σ and the final form ς. Two forms of lowercase S occur in early English manuscripts and printing. Today the medial form has disappeared and the final form has taken over, but both forms were still in common use in Dürer's time and both occur in Dürer's own writing, for example in the banderole of his drawing *Orfe der erst Pusaner.* The prefix to the date of *Adam and Eve* is S. The variant forms of the numerals 5 and 6 in the magic square can be read, respectively, as forms of final s and medial s (approximately, σ). One has the eyes and tongue of a serpent profile, another shows a serpent's face head-on, so each refers to God twice, once as an abbreviation for *salus* and once as the serpent ideogram for God found in the *Hieroglyphica.* The two variant fives and the flourish in the banderole are instances of the capital S. The two variant sixes are σ, which is also a medial S. Dürer's form of S before the date 1504 in *Adam and Eve* is convincingly similar to the sixes in the square. The variant nine in the square is a reflected form of the final S, and deserves special attention. The otherwise mysterious variant of nine after the name 'Albert' in *Adam and Eve* may be seen as the medial S in the lower left-hand corner of the plaque, overturned.

I propose to read all these variants of S as standing for *salus* or salvation, and when they are serpents, for God. The 9 following Albrecht's name in *Adam and Eve* would then be another claim of personal divinity.

The 1 in Dürer's table and the 'I' in the motto are likely synonymous, the § in the motto and the touch of the angel's wing in the table showing its sanctity. The § in the anagram then declares the sanctity of Gateway I to Heaven, natural philosophy, a capital heresy.

We thus cover all ten alphanumeric anomalies in MELENCOLIA§I, plus one in *The Knight* and two in *Adam and Eve* with one hypothesis: they denote the sacred. With the wave and the subliminal faces, they make MELENCOLIA§I pro-science, pro-art, pro-religion, and anti-Church.

As though to verify this interpretation, Jan Wierix in his 1605 version of MELENCOLIA§I systematically modified Dürer's work:

- Wierix eliminated the ambiguous wave.
- He deleted the sacred §.
- He reduced the magnification and elevation of the solitary one in the magic square.

- He flattened the fours and eliminated their folds, so that they are no longer solid crosses but mere numerals.
- He corrected the five within the date in the magic square.
- He corrected the six not touching the date in the magic square.
- He retained the variants of five, six, and nine near the date in the magic square like the S before the date of *The Knight.*

Wierix thus stripped the engraving of six heretical graphic elements and retained two conforming elements. The subliminal versions of the covert message are gone. Even the anagram has become innocuous. If we suppose that he could have eliminated each of these nine features as likely as not, the odds for doing all this by chance is roughly 1 in $2^9 = 1/512$.

Wierix also modified the subliminal faces. He almost eliminated all four ghosts, replaced the quaternity of friends by one cadaverous face, and eliminated First Worship, the skullcap of Second Worship, and First Fool.[1] He left no quaternity or trinity of faces. He also eliminated the ambiguity of the smile of the angel. His angel is unmistakably serious, even sad, consistent with the overt message.

I conclude that probably some form of my interpretation of these nine graphic elements was still alive in 1605 and known to Wierix. These interpretations of the wandering Ss and crosses make sense only if there were viewers who would understand them. Since the doctrines of MELENCOLIA§I are part of the humanist tradition, most likely the initiates were humanists, in particular Dürer's best friend, Willibald Pirckheimer.

3.28 Serpents

After writing the above I realized that I had noticed but not understood the sinuousity of the nine in MELENCOLIA§Ia. Examine the figure and note that it is a rather detailed serpent.

- The top of the nine in the table has a serpent's head curled inside the loop of its body, with two eyes looking directly at us, a mouth-line, and a forked tongue.

In MELENCOLIA§Ia I must cock my head about 30° to the left to best see this face with its two eye-dots. In MELENCOLIA§Ib the nine is reversed, the serpent's face is upright, and the eyes have grown from miniscule dots into discs, all serving to make the serpent more conspicuous.

- The numerals 6 in the Dürer table are also serpents, with head and forked tongue at their lower end. These are the most minute of the subliminal structures. The forked tongues are conveyed by two scratches of the burin.
- The numeral 0 in the Dürer table of MELENCOLIA§I is a serpent eating its tail, the Ourobourous. The eyes are drawn only in MELENCOLIA§Ib.

The Serpent in the nine seemed to refute my then current interpretation. The S that I had deduced was a sign of the highest was the lowest of the low. But within the hour I remembered that I had experienced this inversion before, first with the dog and then with the demon. I went back to Horapollo and looked up ‘snake’. Nothing. Then ‘Serpent’:

> They symbolize the Almighty by the perfect animal, again drawing a complete serpent. Thus among them that which pervades the whole cosmos is Spirit. ([17], Book I, Symbol 64)

So the serpent in the table does not wreck the interpretation but supports it.

There is no clear indication in MELENCOLIA§I that Dürer differed with Christian orthodoxy on the divinity of Jesus; it is a humanist manifesto, not a Unitarian one. The first known Unitarian publication came more than a decade after this engraving.

In this theory, Dürer hides the bell-ringer in obedience to the commandment about graven images but shows the serpents and the eagle because they are not images of God but merely glyphs.

We see that the magic square is riddled with defects, such as inverted, backwards, or otherwise distorted numerals. I suggest that these imperfections were meant to disqualify this engraving for the talisman trade, which still flourished. Not only did Dürer disdain the magic arts, it would have been dangerous to be suspected of practicing them. Most of the distortions inject symbols of God into the magic square, as though to fend off accusations of devil worship.

3.29 The three gnomons

I thought I had exhausted the symbols in MELENCOLIA§I until I noticed that Dürer provided a gnomon for each world. The elemental world has a carpenter’s square, then called a gnomon, lying on the ground. The celestial world has the gnomon magic square. By elimination, the intellectual sphere has the Sun-dial, the highest element of the engraving. The meanings and the locations of the gnomons correlate too well with the three worlds to be easily dismissed as coincidence, nor is there any problem in finding a reasonable guess at his meaning.

4. Summation

There are several main ideas beneath the surface of this engraving.

The motto ‘MELENCOLIA§I’, in its overt form, and the general gloom of the picture, refer to the melancholy of those who strive in vain to define or create absolute truth and beauty. On the surface this is the Agrippan melancholy, not depression but a frenzied and sanctified creativity. In the deeper layer of meaning, it is a despair of ever reaching the absolute truth through our senses. Nevertheless, Dürer tells us, the arts and sciences, including in particular his own graphic arts, are a gateway to Heaven at least as holy as theology, while astrology leads nowhere. Dürer sets natural philosophy, based on the observation and portrayal of Nature in mathematical terms, and theological philosophy, based on the Holy Scriptures, on one level as gateways to Heaven. The creative thinkers of the early 16th century did not separate art and science as neatly as those of the twenty-first; when Dürer sanctifies artistic depiction he sanctifies scientific observation as well. Natural philosophy is not a sure gateway to Heaven, as shown by the various perspectival ambiguities, but neither is theological philosophy, as shown by the three warring sects beneath her dress. And mathematical philosophy, meaning then astrology, is not even in the running: Dürer discards the celestial sphere, the middle world of the Neoplatonic triple cosmos, together with astrology, the mathematical philosophy of Agrippa.

This is a humanist manifesto of the impending Reformation, the scientific revolution, and natural philosophy. Dürer anticipates Descartes' universal doubt based on the existence of optical illusions, and also Descartes' two-fold world of matter (the elemental world of *res extensa*) and mind (the intellectual world of *res extensa*), but he is already beyond Descartes in recognizing that our mathematical models can never exactly match reality, that because perception is the source of all our knowledge, our knowledge is relative, not absolute.

We read these messages in the grand overall structure, the fine details, the ambiguous perspective projections, the optical illusions, the orthographic variants, and the anagram of this engraving. Unscrambled, the motto becomes LIMEN CAELO §I, referring both to natural philosophy as the sacred Gateway I to Heaven, and, as a pun on *caelo*, to Dürer's own graving.

Enough of this message was a capital crime in the time and place of Dürer to account for the concealments in the engraving. Yet they are not mere self-protection, as I assumed at first. The multiple ambiguity does not merely cover the message, it *is* the message:

> The lie is in our understanding, and darkness is so firmly entrenched in our mind that even our groping will fail.

For Dürer and MELENCOLIA§I, the world still has a unique meaning, but God alone knows what it is. MELENCOLIA§I with its pervasive ambiguity, is a postmodern work of art done in the 16th century.

5. Disclosure

I disclose here some of my own still developing views on these topics of Dürer, since they undoubtedly influence my interpretation.

Some of the Neoplatonic cosmology referred to in MELENCOLIA§I lingers on today as the crippling belief that Nature is a mathematical system. This is the mathetic fallacy, the mathematical counterpart of the pathetic fallacy, in which the poet's emotions are projected onto Nature. It persists even though today we understand better the evolutionary process that produced and still produces mathematics and physics. The monism that pervades MELENCOLIA§I survives in physics today as the passion for unity and simplicity, and competes with the hunger for richness and complexity, of which the engraving itself is a specimen. The limits to knowledge illustrated in MELENCOLIA§I agree broadly with writings of the era, such as the *On Learned Ignorance* by Nicolas de Cusa.

But today's limits go deeper. We do not merely doubt our ability to perceive absolute reality, we deny the existence of such an absolute. Today, as for Vico, truth is made, not found. Our predicament is less melancholy than Dürer's. We are not deprived of anything that ever existed, and the loss of mechanical certainty is more than compensated with an increase in understanding and control that was unimaginable before quantum theory.

Descriptive geometry is an important beginning but is still passive. The winged boy still has to learn to value experiments.

Dürer, Kepler, and Einstein wanted to know the creative mind of God, at least metaphorically:

> I want to know how God created this world. I am not interested in this or that phenomenon, in the spectrum of this or that element. I want to know his thoughts; the rest are details. (Attributed to Albert Einstein)

This inspiring metaphor is an insidious form of the mathetic fallacy. It led Kepler to his Platonic-solid model of the Solar System, a thoroughly mistaken theory, and blocked him from appreciating the beauty of the three laws that he found more empirically, and that are still named after him. It led me to set out to read Nature much as I later read MELENCOLIA§I, but less successfully.

When I began my own studies I expected that it would pass through three stages: I would first find or form a universal theory in which I could at least potentially believe, then compute some of its consequences, and finally compare them with experimental data; then I would return to stage one for an improved version. After some decades I could not help noticing that I was still in the first stage. I surmise now that physics is not done to replace one eternal faith by another, but replaces the method of faith by the method of experiment. A final theory makes as little sense as a final experiment.

This is more a source of joy than melancholy. Some physicists still cherish the concept of a single eternal law of Nature or an all-encompassing theory, and the quest for one has stimulated much important and fruitful work. This concept is inconsistent with the quantum theory, which leaves most of the Universe out of the field of view.

It would be simplistic to suppose that such a complex institution as physics serves but one function, but now that the idea of physics as the mechanical representation of Nature has fallen away, truer functions step forward. Physics is both a key element in the survival strategy of the human race, and an outlet for the ancient human need to converse with something larger than ourselves. Nature is among other things an evolutionary process that births us, our purposes, and through us our theories, which are all like millstones that eventually slip off their axles and have to be rolled away.

Acknowledgments

I own the errors in this work. I also owe much to others. The interpretation of MELENCOLIA§I offered would have been impossible without those of Erwin Panofsky and Francis Yates and the studies of Agrippa by Judit Gellérd and by Charles Nauert. I thank Dr Basimah Khulusi and Professor Bill Branch for their interpretations of several elements of MELENCOLIA§I, some of which I ultimately accepted. Khulusi pointed out the reference to Jesus in the four nails of the engraving, and made a convincing model of the octahedron that established the angle 89.5°, among numerous other contributions. Branch pointed out that the numeral 5 in the magic square was upside down and that the 9 was backwards, and brought up the Ouroubouros. Roy Skodnick introduced me to the work of Yates; Heinrich Saller gave me the print of MELENCOLIA§I that triggered this process; Rabbi Mario Karpuj reminded me that Jacob's ladder is the biblical gate of Heaven. Shalom Goldman supplied the reference to Yates on Dürer, the meaning of the Magen David in the Renaissance, and stimulating discussions. Danny Lunsford referred me to NASA on meteors. Glenn Michael referred me to Boorsch [6]; Carla Singer lent me her expertise in art history; Aria Ritz Finkelstein referred me to Velasquez, prepared many of the illustrations, discovered many subliminal faces, and explained their meaning. Shlomit Ritz Finkelstein provided lively discussions, help with Torah and Descartes, and useful expository suggestions. An earlier stage of this study was published in *The St Ann's Review*, whose reviewer improved the paper [11].

Bibliography

[1] Dürer A 2003 *Vnderweysung der Messung mit dem Zirckel un Richtscheyt in Linien, Ebnen und Gantzen Corporen* Vol. 1–4 (Oakland, CA: Octavo) first published 1525, Nuremberg

[2] Panofsky E 1971 *The Life and Art of Albrecht Dürer* (Princeton, NJ: Princeton University Press) The standard work. It reproduces the Dürer works mentioned here

[3] Yates F A 1979 *The Occult Philosophy in the Elizabethan Age* (London: Routledge) see, in particular, chapter 6

[4] Doorly P 2004 Dürer's Melencolia I: Plato's abandoned search for the beautiful *Art Bull.* **88** 255

[5] Agrippa H C 1651 *Of Occult Philosophy* (digital edition, based on the translation of De occulta philosophia libri tres (Three Books About Occult Philosophy) by ed J H Peterson (London: Moule) (http://www.esotericarchives.com/agrippa/agrippa1.htm) vol 1 first published 1531, Paris, vols 2 and 3 1533, Cologne

[6] Boorsch S and Orenstein N M 1997 The print in the North. The age of Albrecht Dürer and Lucas van Leyden *Metropolitan Museum Art Bull.* **54**(4) 13–60

[7] Brant S 1494 *Narrenschi (Ship of Fools)* (1st edition) (New York: Dover) Illustrated by A Dürer, Basel
Barclay A 1509 *Ship of Fools* translator Pynson
Zeydel E H 1944 *Ship of Fools* (rhyming translation) (New York: Columbia University Press); reprinted 1962 (New York: Dover)

[8] Brosseder C 2005 The writing in the Wittenberg sky: astrology in sixteenth century Germany *J. History Ideas* **66** 557–76

[9] Conway W M 1899 *Literary Remains of Albrecht Dürer* (London: University Press) pp 142–3

[10] Federico P J 1972 The melancholy octahedron *Math. Mag.* **45** 30–6

[11] Finkelstein D R 2004 *Melencolia I*: Dürer decoded *St Ann's Rev.*

[12] Galilei G 1953 Dialogue on the Great World Systems (The Salisbury translation revised, annnotated, and with an introduction by G de Santillana) (Chicago, IL: University of Chicago Press) first published 1632, Florence

[13] Gellérd J *Francis Dávid's Epistemological Borrowings from Henry Cornelius Agrippa of Nettesheim* essay (http://unitarius.uw.hu/cffr/papers/agrippa.htm)

[14] Goldman S 2004 *God's Sacred Tongue* (Chapel Hill, CA: University of North Carolina Press)

[15] Greenblatt S 2004 *Will in the World: How Shakespeare Became Shakespeare* (London: Bodley Head)

[16] Heckscher W S 1978 Melancholia (1541). An essay in the rhetoric of description by Joachim *Camerarius Joachim Camerarius* (1500–1574). *Beiträge zur Geschichte des Humanismus im Zeitalter der Reformation* (*Essays on the History of Humanism During the Reformation*) ed Frank Baron (Munich: Humanistische Bibliothek Abhandlungen) pp 32–103

[17] Horapollo N (translator) 1993 *The Hieroglyphics of Horapollo* (Princeton, NJ: Princeton University Press)

[18] Kepler J 1956 *Mysterium cosmographicum de admirabili proportione orbium coelestium* Duncan A M 1981 (translation) *The secret of the universe – Mysterium cosmographicum* introduction and commentary by E J Aiton, with a preface by I Bernard Cohen (New York: Abaris Books)

[19] Dr Basimah Khulusi 2005 personal communication

[20] MacGillavray C 1981 The polyhedron in A Dürer's engraving *Melencolia I Nederl. Akad. Wetensch. Proc.* B **84** 287–94

[21] Nauert C G Jr 1965 *Agrippa and the Crisis of Renaissance Thought* (Urbana, IL: University of Illinois Press)

[22] de Santillana G and von Dechend H 1969 *Hamlet's Mill: An Essay on Myth and the Frame of Time* Argumentation for the astronomical interpretation of all myth (Boston, MA: David R Godine)

[23] Davidson H E and Fisher P (translators) 1979 *Saxo Grammaticus: The History of the Danes* (Cambridge: Cambridge University Press)

[24] Schuster P-K 1991 *MELENCOLIA I: Dürers Denkbild* (Berlin: Gebr. Mann)

[25] Wade W C 1898 *The Symbolisms of Heraldry or A Treatise on the Meanings and Derivations of Armorial Bearings* (London: George Redway)

[26] Weitzel H 2004 A further hypothesis on the polyhedron of A Dürer's engraving Melencolia I *Historia Math.* **31** 11–4

Further reading

Eisler C 1992 Review of Jane Campbell Hutchison *Albrecht Dürer: A Biography Renaissance Quart.* **45** 163–6

Erasmus Desiderius 1991 (reprinted) *In Praise of Folly* (Wilson J: translator) (New York: Dover)

Nicholas of Cusa 1954 *Of Learned Ignorance* (London: Routledge) first published 1440

Penrose R 2004 *The Road to Reality* (New York: Knopf) A brilliant exposition of contemporary Platonism, the antithesis to our thesis

Schreiber P 1999 A new hypothesis on Dürer's enigmatic polyhedron in his copper engraving *Melancholia I. Historia Math.* **26** 369–77

Serveto M 1531 *De Trinitatis Erroribus (On the Errors of the Trinity)* (Hagenau: John Setzer)

Yates F A 1964 *Giordano Bruno and the Hermetic Tradition* (Chicago, IL: University of Chicago Press)

www.ingramcontent.com/pod-product-compliance
Lightning Source LLC
LaVergne TN
LVHW080249110826
845148LV00023BA/875
9781681740263